Worked Examples in Engineering in SI Units

Volume II Electrical Science

The other two volumes

Worked Examples in Engineering in SI Units

Volume II   Electrical Science

Matthew Bates

Lecturer in Electrical Engineering,
Middlesex Polytechnic (Enfield)

London. George Allen & Unwin Ltd
Ruskin House   Museum Street

First published in 1974

ISBN 0 04 620004 5 *hardback*
    0 04 620005 3 *paperback*

Printed in Great Britain
in 10 on 12 pt 'Monophoto' Times Mathematics Series 569
by Page Bros (Norwich) Ltd, Norwich

Foreword

One of the difficulties experienced by students on engineering courses is that the time available for formal instruction is limited. Contact time with lecturers is necessarily devoted to establishing the basic principles of the relevant technology, and too little time is available for the important task of solving problems and obtaining answers.

This book is the second of three volumes, suitable for craft and technician courses, which will allow a student to follow a step-by-step solution to a particular problem and then to solve additional problems provided in the text.

Signs, symbols and abbreviations conform with the latest relevant British Standards, but in these days of change and conversion to SI units it may be possible that some last-minute changes have not been included.

The author would like to thank Mr R. F. Haussmann, B.Sc., Mr A. Fyfe, B.Sc., and other colleagues who have helped in the preparation of this book

Cambridge   N. Hiller

Contents

Chapter 1

Resistive Networks and Circuits

1.1 Kirchhoff's Laws

1. The algebraic sum of the currents at any point in a circuit or network is zero. $\Sigma I = 0$.

   Currents flowing toward the point are taken as being positive and those flowing from the point as negative. This law may also be expressed as 'the sum of the currents flowing towards a point in a circuit or network equals the sum of the currents flowing away from the point'.

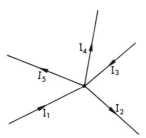

Fig. 1.1 Kirchhoff's first law.

Fig. 1.1 shows a number of current carrying conductors connected to a common point and the current equation may be written as

$$I_1 - I_2 + I_3 - I_4 - I_5 = 0$$

or
$$I_1 + I_3 = I_2 + I_4 + I_5$$

2. In any circuit or closed network the algebraic sum of the e.m.f.s in that circuit or network is equal to the algebraic sum of the voltage (IR) drops in that circuit or network.

   Fig. 1.2 shows such a circuit in which current directions have been assumed and these fix the directions in which the voltage drops occur.

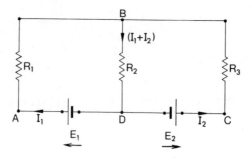

Fig. 1.2 Kirchhoff's second law.

In circuit *ABDA*,

$$E_1 = I_1 R_1 + R_2 (I_1 + I_2)$$
$$= I_1 (R_1 + R_2) + I_2 R_2$$

In circuit *ABCDA*,

$$E_1 - E_2 = I_1 R_1 - I_2 R_3$$

Note: since E_1 is assumed to be greater than E_2 then the voltage drop across R_3 is taken to be negative as it is in the opposite direction to the current I_2. Any IR drop taken in the direction of the current flow is positive and any IR drop taken in the direction opposite to the current flow is negative, i.e. it is a voltage rise.

Example 1.1
Fig. 1.3 shows a system of batteries and resistors. By applying Kirchhoff's laws determine the value of current in each resistor.

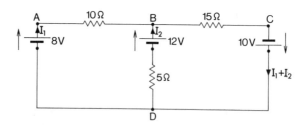

Fig. 1.3 Circuit for Example 1.1.

Let the current directions be as shown in Fig. 1.3. Consider *DBAD*

$$12 - 8 = 5I_2 - 10I_1$$
$$I_1 = \frac{5I_2 - 4}{10} A$$

Consider *DBCD*.

$$12 + 10 = 5I_2 + 15(I_1 + I_2)$$
$$22 = 15I_1 + 20I_2$$

Substitute for I_1.

$$22 = 15\left(\frac{5I_2 - 4}{10}\right) + 20I_2$$
$$= \frac{15I_2 - 12}{2} + 20I_2$$
$$44 = 15I_2 - 12 + 40I_2$$
$$56 = 55I_2$$
$$I_2 = 1\tfrac{1}{55}\,\text{A} = 1{\cdot}018\,\text{A}$$
$$I_1 = (5\tfrac{1}{11} - 4)/10$$
$$= \tfrac{12}{110}\,\text{A} = 0{\cdot}109\,\text{A}$$

Note: It is often more convenient and more accurate to use the vulgar fraction in substitution than the decimal fraction.

$$I_1 + I_2 = 1\tfrac{7}{55}\,\text{A} = 1{\cdot}127\,\text{A}$$

The currents are in the directions shown in Fig. 1.3.
    The current in branch *AB* is 0·109 A.
    The current in branch *DB* is 1·018 A.
    The current in branch *BC* is 1·127 A.
    (Answers are given to instrument accuracy.) The accuracy of the answers may easily be checked.

In circuit *DBAD*,

$$12 - 8 = (5 \times 1\tfrac{1}{55}) - (10 \times \tfrac{6}{55})$$
$$= 5\tfrac{5}{55} - 1\tfrac{5}{55} = 4\,\text{V}$$

Alternative solution. Fig. 1.4 shows the same circuit as in Fig. 1.3 but now it is

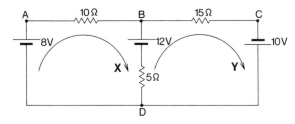

Fig. 1.4 Circuit for alternative solution to Example 1.1.

assumed that circulating currents X and Y flow round the meshes as shown. Consider $ABDA$ in Fig. 1.4

$$8 - 12 = 10X + 5(X - Y)$$
$$-4 = 15X - 5Y.$$

Consider $DBCD$.

$$12 + 10 = 15Y + 5(Y - X)$$
$$22 = 20Y - 5X$$
$$5X = 20Y - 22$$

From mesh $ABDA$,

$$15X = 5Y - 4$$

Hence

$$60Y - 66 = 5Y - 4$$
$$55Y = 62$$
$$Y = \tfrac{62}{55} = 1\tfrac{7}{55}$$
$$= 1 \cdot 127 \text{ A} \quad \text{(current in 15 } \Omega \text{ resistor)}$$
$$15X = \frac{62 \times 5}{55} - 4$$
$$= \frac{62 - 44}{11} = \frac{18}{11}$$
$$X = \tfrac{18}{11} \times \tfrac{1}{15} = \tfrac{6}{55} \text{ A}$$
$$= 0 \cdot 109 \text{ A} \quad \text{(current in 10 } \Omega \text{ resistor)}$$
$$Y - X = 1 \cdot 127 - 0 \cdot 109$$
$$= 1 \cdot 108 \text{ A} \quad \text{(current in 5 } \Omega \text{ resistor)}$$

Example 1.2

Two batteries of e.m.f. 50 V and 45 V, each have an internal resistance of 1 Ω. If the batteries are connected in parallel to supply a 10 Ω resistor, determine the current supplied by each battery.

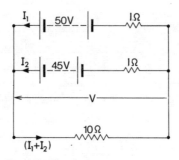

Fig. 1.5 Circuit for Example 1.3.

In the circuit diagram shown in Fig. 1.5 the internal resistances are shown, for convenience, as series resistors. Let V be the voltage across the combination and let the current distribution be as shown. Then

$$V = 50 - I_1 \times 1$$

and

$$V = 45 - I_2 \times 1$$

Hence

$$50 - I_1 = 45 - I_2$$

$$I_1 = 5 + I_2$$

Also

$$V = 10(I_1 + I_2)$$

$$45 - I_2 = 10[(5 + I_2) + I_2]$$

$$45 - I_2 = 50 + 20I_2$$

$$-5 = 21I_2$$

$$I_2 = \frac{-5}{21} = 0 \cdot 238 \text{ A}$$

in the direction opposite to that shown in Fig. 1.5.
Substituting for I_2,

$$I_1 = 5 - \tfrac{5}{21} = 4\tfrac{16}{21}$$

$$= 4 \cdot 763 \text{ A}$$

in the direction as shown in Fig. 1.5.
    Check. Since

$$50 - I_1 = 45 - I_2$$

hence

$$50 - 4\tfrac{16}{21} = 45 - (-\tfrac{5}{21})$$

$$45\tfrac{5}{21} = 45\tfrac{5}{21}.$$

Example 1.3
Determine the current in each branch of the circuit shown in Fig. 1.3, if the 10 V battery were reversed.

Example 1.4
Two batteries A and B are connected in parallel to supply a 5 Ω resistor. The e.m.f.s of A and B are 100 V and 110 V and their internal resistances are 1 Ω and 1·5 Ω respectively. Determine the battery currents.

Example 1.5
An unknown resistance is connected to a supply of two batteries in parallel. One battery, having an e.m.f. of 90 V and an internal resistance of $1\tfrac{2}{3} \Omega$ supplies 6 A and the second battery has an e.m.f. of 85 V and an internal resistance of 1·25 Ω. Calculate the current supplied by the second battery and the value of the load resistance.

Example 1.6
Fig. 1.6 shows a circuit diagram. Determine the magnitude and directions

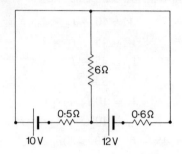

Fig. 1.6 Circuit for Example 1.6.

of the currents in each part of the circuit. If the 10 V battery were reversed what then would be the current in the 6 Ω resistor?

Example 1.7
A battery, having an e.m.f. of 20 V and an internal resistance of 0·6 Ω, is connected across a 10 Ω resistor. A second battery, having an e.m.f. of 30 V and an internal resistance of 1 Ω, is connected in parallel with the first battery but a 7 Ω resistor is connected between the positive terminals of the two batteries. Calculate the currents in each part of the circuit.

Example 1.8
Two batteries *A* and *B* are connected in parallel and are connected through a 12 Ω resistor to a charging voltage of 100 V. The e.m.f.s and internal resistances of the batteries are 90 V and 85 V, and 0·3 Ω and 0·25 Ω respectively. Determine the current in each battery and the total charging current.

Example 1.9
A battery, having an e.m.f. of 9 V and an internal resistance of 1·2 Ω, is made up of similar cells. A resistor, having a resistance of 2·6 Ω is connected across the battery while a second resistor of 1·2 Ω resistance is connected between the negative terminal of the battery and a point in the battery which is at a potential of 6 V above the potential of the negative terminal. Calculate the currents in the two resistors.

1.2 Distribution Circuits

Example 1.10
A two-wire distributor cable is 1·5 km long and supplies loads of 50 A, 100 A, and 125 A, situated at distances of 500 m, 800 m and 1200 m from the feeding

end which is maintained at 500 V. Calculate the p.d. at each load point when each conductor of the cable has a resistance of 0·05 Ω/km.

Fig. 1.7 shows the circuit for the system. Since the system is fed at one end

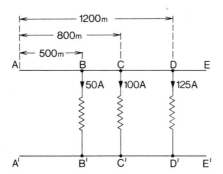

Fig. 1.7 Circuit for twin feeder fed at one end—Example 1.10.

only the total current input to the circuit must be the sum of the load currents. Therefore

$$\text{Current input} = 50 + 100 + 125\,\text{A} = 275\,\text{A}$$

Applying Kirchhoff's First Law the current distribution will be

$$\text{Current } A \text{ to } B = 275\,\text{A} = \text{Current } B' \text{ to } A'$$
$$\text{Current } B \text{ to } C = 275 - 50 = 225\,\text{A}$$
$$= \text{Current } C' \text{ to } B'$$
$$\text{Current } C \text{ to } D = 225 - 100 = 125\,\text{A}$$
$$= \text{Current } D' \text{ to } C'$$

Since the two wires of each section carry the same current they are in series and so their resistances may be added.

Length of AB = 500 m

Length of BC = 800 − 500 = 300 m

Length of CD = 1200 − 800 = 400 m

Resistance of AB $= 2 \times \dfrac{500}{1000} \times 0{\cdot}05 = 0{\cdot}05\,\Omega$

Resistance of BC $= 2 \times \dfrac{300}{1000} \times 0{\cdot}05 = 0{\cdot}03\,\Omega$

Resistance of CD $= 2 \times \dfrac{400}{1000} \times 0{\cdot}05 = 0{\cdot}04\,\Omega$

Total IR drop A to $B = 275 \times 0{\cdot}05 = 13{\cdot}75\,\text{V}$

17

N

P.D. at B $= 500 - 13{\cdot}75 = 486{\cdot}25$ V

Total IR drop B to $C = 225 \times 0{\cdot}03 = 6{\cdot}75$ V

P.D. at C $= 486{\cdot}25 - 6{\cdot}75 = 479{\cdot}50$ V

Total IR drop C to $D = 125 \times 0{\cdot}04 = 5{\cdot}00$ V

P.D. at D $= 479{\cdot}5 - 5{\cdot}0 = 474{\cdot}50$ V

Potential differences at the load points are 486 V, 479 V and 474 V.

Example 1.11
A ring main distributor is 500 m long and is supplied with 240 V at a point AA'. One load, which takes a current of 100 A, is connected at a point 200 m from AA' and a second load, which takes a current of 200 A, is connected 150 m from AA' in the opposite direction. The resistance of 100 m of single

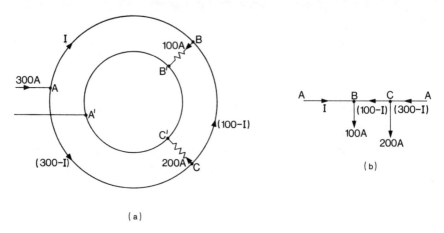

(a)

Fig. 1.8 (a) Circuit of ring main for Example 1.11.
         (b) Equivalent two-wire feeder fed at both ends at the same potentials.

conductor is 0·02 Ω. Calculate the current in each section of the distributor and the p.d.s at the load points.

Fig. 1.8(a) shows the distributor, the position of the loads at B and at C and the assumed current distribution, applying Kirchhoff's First Law. It is easier to consider the system as being a two-wire distributor fed at both ends at the same potential.

Fig. 1.8(b) represents such a system with the two resistances of each section added together.

Length of AB $= 200$ m

$$\text{Total resistance of } AB = 2 \times 200 \times \frac{0 \cdot 02}{100} \Omega = 0 \cdot 08 \, \Omega$$

Length of BC $\qquad = 150 \, \text{m}$

$$\text{Total resistance of } BC = 2 \times 150 \times \frac{0 \cdot 02}{100} \Omega = 0 \cdot 06 \, \Omega$$

Length of AC $\qquad = 150 \, \text{m}$

Total resistance of $AC = 0 \cdot 06 \, \Omega$

Since the distributor is at the same potential at both ends then the total volt drop along the distributor must be zero. Applying Kirchhoff's Second Law,

$$\begin{aligned}
0 &= 0 \cdot 08 I - 0 \cdot 06(100 - I) - 0 \cdot 06(300 - I) \\
&= 0 \cdot 08 I - 6 + 0 \cdot 06 I - 18 + 0 \cdot 06 I \\
&= 0 \cdot 2 I - 24 \\
I &= 24/0 \cdot 2 = 120
\end{aligned}$$

The current between A and B is 120 A, between B and C it is 20 A and between A and C it is 180 A.

$$
\begin{aligned}
&IR \text{ drop } A \text{ to } B && = 120 \times 0 \cdot 08 = 9 \cdot 6 \, \text{V} \\
&\text{Potential across load at } B = 240 - 9 \cdot 6 = 230 \cdot 4 \, \text{V} \\
&IR \text{ drop } A \text{ to } C && = 180 \times 0 \cdot 06 = 10 \cdot 8 \, \text{V} \\
&\text{Potential across load at } C = 240 - 10 \cdot 8 = 229 \cdot 2 \, \text{V}
\end{aligned}
$$

Alternatively,

$$
\begin{aligned}
&IR \text{ drop } B \text{ to } C && = (120 - 100) \times 0 \cdot 06 = 1 \cdot 2 \, \text{V} \\
&\text{Potential across load at } C = 230 \cdot 4 - 1 \cdot 2 = 229 \cdot 2 \, \text{V}
\end{aligned}
$$

The potential differences across the loads will be 230 V and 229 V.

(Note that with a ring main the p.d.s across the loads are more nearly equal than when a distributor is fed at one end only.)

Example 1.12
A two-wire distributor AB is 2 km long and each conductor has a resistance of $0 \cdot 1 \, \Omega$ per km. The system is fed at 255 V at end A and at 250 V at end B.

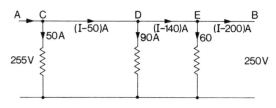

Fig. 1.9 Circuit for Example 1.12.

19

Loads at distances of 200 m, 1000 m and 1500 m take currents of 50 A, 90 A and 60 A respectively. Determine the current in each section of the cable and the p.d. across each load.

Fig. 1.9 shows the distributor, the loads and the assumed current distribution. Since end A is at a higher potential than end B the total volt drop from A to B must equal 5 V.

$$\text{Total resistance } A \text{ to } C = 2 \times 200 \times \frac{0{\cdot}1}{1000} = 0{\cdot}04\,\Omega$$

$$\text{Total resistance } C \text{ to } D = 2 \times 800 \times \frac{0{\cdot}1}{1000} = 0{\cdot}16\,\Omega$$

$$\text{Total resistance } D \text{ to } E = 2 \times 500 \times \frac{0{\cdot}1}{1000} = 0{\cdot}10\,\Omega$$

$$\text{Total resistance } E \text{ to } B = 2 \times 500 \times \frac{0{\cdot}1}{1000} = 0{\cdot}10\,\Omega$$

Applying Kirchhoff's Second Law,

$$5 = 0{\cdot}04I + 0{\cdot}16(I - 50) + 0{\cdot}1(I - 140) + 0{\cdot}1(I - 200)$$
$$= 0{\cdot}04I + 0{\cdot}16I - 8 + 0{\cdot}1I - 14 + 0{\cdot}1I - 20 = 0{\cdot}4I - 42$$
$$47 = 0{\cdot}4I$$
$$I = 117{\cdot}5\,\text{A}$$

Current distribution:

Current in $AC = 117{\cdot}5$ A, Current in $CD = 67{\cdot}5$ A, Current in $DE = -22{\cdot}5$ A, Current in $EB = -82{\cdot}5$ A.

$$\text{Total } IR \text{ drop } A \text{ to } C = 117{\cdot}5 \times 0{\cdot}04 = 4{\cdot}7\,\text{V}$$
$$\text{P.D. across load at } C = 255 - 4{\cdot}7 = 250{\cdot}3\,\text{V}$$
$$\text{Total } IR \text{ drop } C \text{ to } D = 67{\cdot}5 \times 0{\cdot}16 = 10{\cdot}8\,\text{V}$$
$$\text{P.D. across load at } D = 250{\cdot}3 - 10{\cdot}8 = 239{\cdot}5\,\text{V}$$
$$\text{Total } IR \text{ drop } D \text{ to } E = -22{\cdot}5 \times 0{\cdot}1 = -2{\cdot}25\,\text{V}$$
$$\text{P.D. across load at } E = 239{\cdot}5 + 2{\cdot}25 = 241{\cdot}75\,\text{V}$$
$$\text{Total } IR \text{ drop } E \text{ to } B = -82{\cdot}5 \times 0{\cdot}1 = -8{\cdot}25\,\text{V}$$
$$\text{P.D. at } B \qquad = 241{\cdot}75 + 8{\cdot}25 = 250\,\text{V}$$

The potential differences across the loads will be 250 V, 240 V and 242 V.

Note: The negative currents in DE and EB indicate that the true direction of current is from B to E and from E to D. The negative voltage drops from D to E and from E to B indicate that in fact the voltage rises.

Example 1.13

In a section of a 3-wire, d.c. system, 1 km long and fed from both ends at the same potential, there are loads of 30 A and 60 A connected between the positive and the neutral at distances of 300 m and 750 m from one end respectively. On the negative side there are loads of 40 A at 250 m and 30 A at 800 m from the same end. Each outer has a resistance of 0·5 Ω and the neutral has half the cross-sectional area of either outer. If a voltage of 240 V is maintained between the neutral and each outer calculate the p.d. across each load.

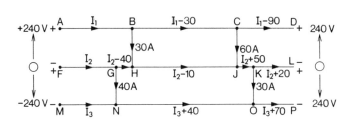

Fig. 1.10 Circuit of three wire distributor in Example 1.13.

Fig. 1.10 shows the 3-wire system and the assumed current distribution, applying Kirchhoff's First Law. Consider the positive feeder *AD*.

$$\text{Resistance of section } AB = \frac{0\cdot5}{1000} \times 300 = 0\cdot15 \ \Omega$$

$$\text{Resistance of section } BC = \frac{0\cdot5}{1000} \times 450 = 0\cdot225 \ \Omega$$

$$\text{Resistance of section } CD = \frac{0\cdot5}{1000} \times 250 = 0\cdot125 \ \Omega$$

Since the potentials at each end of the feeder are the same, then the total volt-drop along the cable must be zero. Hence

$$0\cdot15I_1 + 0\cdot225(I_1 - 30) + 0\cdot125(I_1 - 90) = 0$$
$$0\cdot15I_1 + 0\cdot225I_1 - 6\cdot75 + 0\cdot125I_1 - 11\cdot25 = 0$$
$$0\cdot5I_1 - 18 = 0$$
$$0\cdot5I_1 = 18$$
$$I_1 = 36 \ \text{A}.$$

Therefore $_AI_B = 36 \ \text{A}$: $_BI_C = 6 \ \text{A}$: $_DI_C = 54 \ \text{A}$.

Consider the negative feeder *MP*.

$$\text{Resistance of section } MN = \frac{0\cdot5}{1000} \times 250 = 0\cdot125 \ \Omega$$

21

$$\text{Resistance of section } NO \ = \frac{0{\cdot}5}{1000} \times 550 = 0{\cdot}275 \ \Omega$$

$$\text{Resistance of section } OP \ = \frac{0{\cdot}5}{1000} \times 200 = 0{\cdot}1 \ \Omega$$

$$0{\cdot}125I_3 + 0{\cdot}275(I_3 + 40) + 0{\cdot}1(I_3 + 70) = 0$$
$$0{\cdot}125I_3 + 0{\cdot}275I_3 + 11 + 0{\cdot}1I_3 + 7 = 0$$
$$0{\cdot}5I_3 + 18 = 0$$
$$0{\cdot}5I_3 = -18$$
$$I_3 = -36 \ \text{A}$$

Hence $_NI_M = 36$ A; $_NI_O = 4$ A; $_OI_P = 34$ A.

Consider the neutral feeder FL. Since the neutral has only half the cross-sectional area of either outer then the resistance of the neutral is twice that of either outer. ($R \propto 1/\text{area}$).

$$\text{Resistance of 1 km of neutral} = 1 \ \Omega$$

$$\text{Resistance of section } FG \ = \frac{1}{1000} \times 250 = 0{\cdot}25 \ \Omega$$

$$\text{Resistance of section } GH \ = \frac{1}{1000} \times \ \ 50 = 0{\cdot}05 \ \Omega$$

$$\text{Resistance of section } HJ \ = \frac{1}{1000} \times 450 = 0{\cdot}45 \ \Omega$$

$$\text{Resistance of section } JK \ = \frac{1}{1000} \times \ \ 50 = 0{\cdot}05 \ \Omega$$

$$\text{Resistance of section } KL \ = \frac{1}{1000} \times 200 = 0{\cdot}2 \ \Omega$$

Then,

$$0{\cdot}25I_2 + 0{\cdot}05(I_2 - 40) + 0{\cdot}45(I_2 - 10) + 0{\cdot}05(I_2 + 50) + 0{\cdot}2(I_2 + 20) = 0$$
$$0{\cdot}25I_2 + 0{\cdot}05I_2 - 2 + 0{\cdot}45I_2 - 4{\cdot}5 + 0{\cdot}05I_2 + 2{\cdot}5 + 0{\cdot}2I_2 + 4 = 0$$
$$I_2 = 0 \ \text{A}$$

Hence $_FI_G = 0$ A; $_GI_H = 40$ A; $_HI_J = 10$ A; $_JI_K = 50$ A; $_LI_K = 20$ A.

When the current distribution has been determined the potential at each point may be found as follows.

$$\text{Volt drop from } A \text{ to } B \ = 0{\cdot}15 \times 36 = 5{\cdot}4 \ \text{V}$$
$$\text{Potential at } B \qquad\quad = 240 - 5{\cdot}4 = 234{\cdot}6 \ \text{V}$$
$$\text{Volt drop from } B \text{ to } C \ = 0{\cdot}255 \times 6 = 1{\cdot}35 \ \text{V}$$

Potential at C $\quad= 234{\cdot}6 - 1{\cdot}35 = 233{\cdot}25$ V

Volt drop from F to G $\quad= 0{\cdot}25 \times 0 = 0$ V

Potential at G $\quad= 0$ V

Volt drop from G to H $= 0{\cdot}05 \times -40 = -2$ V

Potential at H $\quad= 0 - (-2) = 2$ V

Volt drop from H to J $= 0{\cdot}45 \times -10 = -4{\cdot}5$ V

Potential at J $\quad= 2 - (-4{\cdot}5) = 6{\cdot}5$ V

Volt drop from J to K $= 0{\cdot}05 \times 50 = 2{\cdot}5$ V

Potential at K $\quad= 6{\cdot}5 - 2{\cdot}5 = 4$ V

Volt drop from M to N $= 0{\cdot}125 \times -36 = -4{\cdot}5$ V

Potential at N $\quad= -240 - (-4{\cdot}5) = -235{\cdot}5$ V

Volt drop from N to O $= 0{\cdot}275 \times 4 = 1{\cdot}1$ V

Potential at O $\quad= -235{\cdot}5 - 1{\cdot}1 = -236{\cdot}6$ V

The p.d.s across the various loads may now be determined from the potentials of the relevant points:

p.d. across $BH = 234{\cdot}6 - 2 = 232{\cdot}6$ V

p.d. across $CJ = 233{\cdot}25 - 6{\cdot}5 = 226{\cdot}75$ V

p.d. across $GN = 0 - (-235{\cdot}5) = 235{\cdot}5$ V

p.d. across $KO = 4 - (-236{\cdot}6) = 240{\cdot}6$ V.

The answer should not be given to a greater degree of accuracy than that given by a voltmeter, hence the potential differences across the loads are 233 V; 227 V; 236 V and 241 V.

Example 1.14
A two-wire distributor is 2 km long and each conductor has a resistance of 0·05 Ω. It is supplied at one end at 240 V, and at distances of 500 m, 1200 m and 1500 m from the feeding end are connected loads which take currents of 50 A, 70 A and 80 A respectively. Calculate the current in each section and the p.d. across each load.

Example 1.15
A two core cable is 1·5 km long and supplies loads of 50 A, 100 A and 150 A. If the cable is fed at one end at 240 V estimate the greatest distance the load taking 150 A may be connected from the feeding end, if the p.d. across this load has not to fall below 205 V. Each core of the cable has a resistance of 0·05 Ω per 1000 m and the loads of 50 A and 100 A are at distances of 500 m and 1200 m respectively from the feeding end.

23

Example 1.16
If in Example 1.15 the load taking 150 A were connected as required and the cable were fed at both ends at 240 V, what then would be the p.d.s across the loads?

Example 1.17
A ring main is 350 m long and is supplied at 240 V. A load, connected at a distance of 150 m from the supply, takes a current of 75 A and a second load which takes a current of 100 A, is connected 100 m from the supply point, but in the opposite direction around the ring. If each conductor has a resistance of 0·07 Ω, determine the current in each section of the ring main and the voltage across each of the loads.

Example 1.18
The p.d. at each end of a two wire distributor is maintained at 250 V and loads are connected at distances of 250 m and 700 m from the same end. If the nearer load takes a current of 60 A estimate the greatest current the second load may take if the p.d. across this load must not fall below 246 V. The length of the distributor is 1 km and the resistance of each wire is 0·1 Ω.

Example 1.19
The voltages at the ends of a 750 m section of a distribution system are maintained at 240 V and 242 V. Loads, which take currents of 30 A and 50 A, are connected at distances of 200 m and 450 m respectively from the end at the lower potential. If each conductor has a resistance of 0·4 Ω, calculate the p.d.s across each load.

Example 1.20
A 3-wire distributor *AB*, is 750 m long and is fed at both ends. The voltage between each outer and neutral being 240 V at both *A* and *B*. Between the positive and neutral are loads which take currents of 20 A and 40 A and are connected at distances of 200 m and 500 m respectively from *A*. At a distance of 400 m from *A* a load which takes a current of 30 A is connected between the negative and the neutral. If each outer has a resistance of 0·05 Ω/100 m and the resistance of the neutral is twice that of either outer, determine the p.d. across each of the loads.

Example 1.21
A 3-wire distributor, 1 km in length, is fed at the ends *A* and *B* with 240 V and 242 V respectively between the neutral and each outer. At a distance of 600 m from *A* there is a load between the positive feeder and the neutral which takes a current of 50 A. Between the negative feeder and the neutral there are loads, at distances of 400 m and 800 m from *A*, which take currents of 30 A and 40 A respectively. Determine the current in each section and the

p.d. across each load. Each outer has a resistance of 0·5 Ω and the neutral is half the cross-sectional area of either outer.

1.3 The Superposition Principle

In a circuit, containing a number of sources (e.m.f.s) and resistors, the current through any resistor in the circuit is the algebraic sum of the currents through that resistor when the circuit is considered to contain only one source; each source being considered in turn.

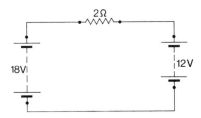

Fig. 1.11 Circuit to illustrate the Superposition Principle.

Fig. 1.11 shows a simple circuit in which it is necessary to determine the current in the 2 Ω resistor. It is obvious that

$$\text{current } I = \frac{18 - 12}{2} = \frac{6}{2} = 3\,\text{A.}$$

Using the Superposition Principle the total current is the algebraic sum of the currents due to each e.m.f. taken separately.

Current due to 18 V source $= \dfrac{18}{2} = 9$ A in clockwise direction

Current due to 12 V source $= \dfrac{12}{2} = 6$ A in anticlockwise direction

Total current $= 9 - 6\,\text{A} = 3\,\text{A}$ in clockwise direction

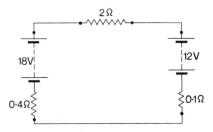

Fig. 1.12 Circuit to illustrate the Superposition Principle.

25

If the batteries had had internal resistances then these resistances must be taken into account for each part of the calculation.

Fig. 1.12 shows the complete circuit with internal resistances included.

Since all the resistances are in series the total resistance is $0.4 + 2.0 + 0.1$, i.e. $2.5\ \Omega$. Neglecting the 12 V source,

$$I = \frac{18}{2.5}\ A = 7.2\ A\ \text{(clockwise)}$$

Neglecting the 18 V source,

$$I = \frac{12}{2.5} = 4.8\ A\ \text{(anticlockwise)}$$

Total current in the $2\ \Omega$ resistor $= 7.2 - 4.8 = 2.4\ A$

$$(\text{compare } (18 - 12)/2.5)$$

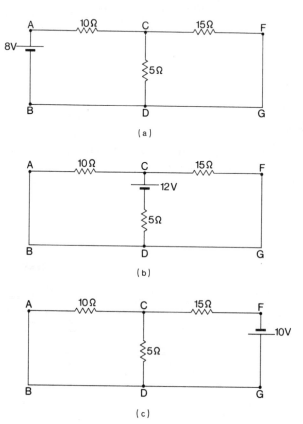

Fig. 1.13 Circuits for the application of the Superposition Principle to Example 1.22.

Example 1.22

Fig. 1.3 shows a system of batteries and resistors. Using the Superposition Principle determine the current in each resistor.

Fig. 1.13(a) shows the circuit when only the 8 V source is considered.

$$\text{Total resistance} = 10 + \frac{5 \times 15}{5 + 15} = \frac{55}{4}\,\Omega$$

$$\text{Total current} = 8 \times \frac{4}{55} = \frac{32}{55} = {}_BI'_A$$

$$_CI'_D = \frac{32}{55} \times \frac{15}{20} = \frac{24}{55}\,\text{A}$$

$$_FI'_G = \frac{32}{55} \times \frac{5}{20} = \frac{8}{55}\,\text{A}$$

Fig. 1.13(b) shows the circuit when only the 12 V source is considered.

$$\text{Total resistance} = 5 + \frac{10 \times 15}{10 + 15} = 11\,\Omega$$

$$\text{Total current} = \frac{12}{11}\,\text{A} = {}_DI''_C = - {}_CI''_D$$

$$_AI''_B = - {}_BI''_A = \frac{12}{11} \times \frac{15}{25} = \frac{36}{55}\,\text{A}$$

$$_FI''_G = \frac{12}{11} \times \frac{10}{25} = \frac{24}{55}\,\text{A}$$

Fig. 1.13(c) shows the circuit when only the 10 V source is considered.

$$\text{Total resistance} = 15 + \frac{5 \times 10}{5 + 10} = \frac{55}{3}\,\Omega$$

$$\text{Total current} = 10 \times \frac{3}{55} = \frac{6}{11} = {}_FI'''_G$$

$$_DI'''_C = - {}_CI'''_D = \frac{6}{11} \times \frac{10}{15} = \frac{4}{11}\,\text{A}$$

$$_BI'''_A = \frac{6}{11} \times \frac{5}{15} = \frac{2}{11}\,\text{A}$$

$$\text{Total current in } BA = \frac{32}{55} - \frac{36}{55} + \frac{2}{11}$$

$$= \frac{42 - 36}{55} = \frac{6}{55}\,\text{A} = 0.109\,\text{A}$$

$$\text{Total current in } CD = \frac{24}{55} - \frac{12}{11} - \frac{4}{11} = \frac{24 - 60 - 20}{55}$$

$$= -\frac{56}{55}A = 1{\cdot}018\,A \text{ in the direction } D \text{ to } C.$$

$$\text{Total current in } FG = \frac{8}{55} + \frac{24}{55} + \frac{6}{11} = \frac{8 + 24 + 30}{55}$$

$$= \frac{62}{55}A = 1{\cdot}127\,A.$$

Compare these answers with those to Example 1.1.

Example 1.23

Two batteries, *A* and *B*, are connected, to supply current, to a load having a resistance of 4 Ω. The details of the batteries are as follows:

Battery *A*: e.m.f. = 30 V and internal resistance of 2 Ω.

Battery *B*: e.m.f. = 25 V and internal resistance of 1 Ω.

Using the Superposition Principle, determine the current taken by the load and that supplied by each battery.

Fig. 1.14 shows the circuit diagram for the example. Fig. 1.15(a) shows the

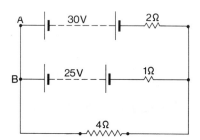

Fig. 1.14 Circuit diagram for Example 1.23.

circuit with battery *B* omitted. Note that when a battery or other source is removed from the circuit it is replaced by a resistor having a resistance equal to the internal resistance of the source removed.

$$\text{Total resistance} = 2 + \frac{4 \times 1}{4 + 1} = 2{\cdot}8\,\Omega$$

$$\text{Total current} = \frac{30}{2{\cdot}8}\,A = I'_A$$

$$\text{Load current} = \frac{30}{2{\cdot}8} \times \frac{1}{5} = \frac{6}{2{\cdot}8}\,A$$

$$\text{Current through resistance of battery } B = \frac{30}{2{\cdot}8} \times \frac{4}{5} = \frac{24}{2{\cdot}8}\,A$$

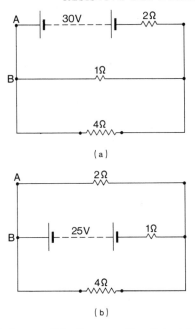

Fig. 1.15 Subcircuits for use of the Superposition Principle in Example 1.23.

Fig. 1.15(b) shows the circuit with battery A replaced by a resistor having a resistance equal to the internal resistance of A.

Total resistance $\qquad = 1 + \dfrac{4 \times 2}{4 + 2} = \dfrac{7}{3}\,\Omega$

Total current $\qquad = 25 \times \dfrac{3}{7} = \dfrac{75}{7}\,\text{A} = I_B''$

Load current $\qquad = \dfrac{75}{7} \times \dfrac{2}{6} = \dfrac{25}{7}\,\text{A}$

Current through resistance of battery $A = \dfrac{75}{7} \times \dfrac{4}{6} = -\dfrac{50}{7}\,\text{A}$

Total load current $\qquad = \dfrac{6}{2 \cdot 8} + \dfrac{25}{7} = \dfrac{15 + 25}{7}$

$\qquad = \dfrac{40}{7} = 5 \cdot 71\,\text{A}$

Current supplied by A $\qquad = \dfrac{30}{2 \cdot 8} - \dfrac{50}{7} = \dfrac{75 - 50}{7}$

$\qquad = \dfrac{25}{7} = 3 \cdot 57\,\text{A}$

Current supplied by B

$$= \frac{75}{7} - \frac{24}{2 \cdot 8} = \frac{75 - 60}{7}$$

$$= \frac{15}{7} = 2 \cdot 14 \, A$$

Alternatively,

Current supplied by B $\qquad = 5 \cdot 71 - 3 \cdot 57 = 2 \cdot 14 \, A$

Example 1.24
Using the Superposition Principle determine the currents in each branch of

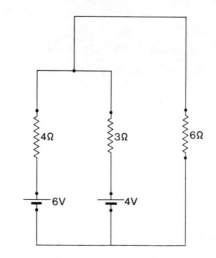

Fig. 1.16 Circuit for Example 1.24.

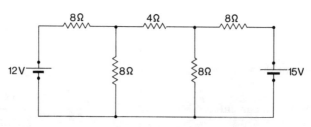

Fig. 1.17 Circuit for Example 1.24.

the circuits shown in Figs. 1.16 and 1.17. Check your answers by applying Kirchhoff's laws.

Example 1.25
Two batteries, having e.m.f.s of 10 V and 12 V and internal resistances of

30

1·0 Ω and 1·2 Ω respectively, are connected in parallel to a load, having a resistance of 3 Ω. Apply the Superposition Principle to calculate the currents supplied by each battery and the load current.

Chapter 2

Wheatstone's Bridge and the Potentiometer

2.1 Wheatstone's Bridge

Wheatstone's Bridge and its adaptations—the metre wire bridge, etc.—are used for measuring resistance. Fig. 2.1 shows the connections for the bridge.

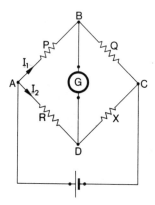

Fig. 2.1 Circuit for Wheatstone's Bridge.

The resistors P and Q are of known value and are usually in some simple ratio, e.g. $1:10$ or $10:1$. The resistor R is a variable resistance, the value of which is known and X is the unknown resistance. The value of R is varied until no current flows in the galvanometer. (Note that a galvo shunt is used to protect the instrument before balance is obtained.)

With no current in the galvanometer circuit points B and D must be at the same potential. Resistors P and R are at the same potential at A, and therefore

$$\text{Volt drop from } A \text{ to } B = \text{Volt drop from } A \text{ to } D$$

$$I_1 \cdot P = I_2 \cdot R \tag{1}$$

Since Q and X are at the same potential at C, then

$$I_1 . Q = I_2 . X \qquad (2)$$

Note that since there is no current in the galvo circuit the resistors P and Q are in series, as are R and X. Divide (2) by (1)

$$I_1 . Q / I_1 . P = I_2 . X / I_2 . R$$
$$Q / P = X / R$$

or

$$X = R \times Q / P$$

It follows from the above that $P \times X = Q \times R$, i.e. the products of opposite resistances are equal. This is the general expression for balanced conditions in a bridge, and it is the best one to remember as it is independent of the letters allocated to the various resistors and their method of connection.

Example 2.1
A Wheatstone's bridge network consists of four resistors, $AB = 10\,\Omega$, $BC = 2\,\Omega$, $AD = 15\,\Omega$, and $DC = 2\Omega$. What alteration must be made to the resistor DC so that no current will flow in a $7\,\Omega$ resistor connected between B and D? If the current taken from the supply connected across AC is $1\cdot5$ A, calculate the current through each resistor after the alteration has been made.

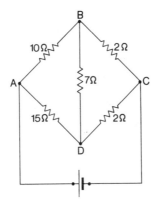

Fig. 2.2 Circuit for Example 2.1.

Fig. 2.2 shows the resistance network. Since no current is to flow in the $7\,\Omega$ resistor, the bridge must be balanced. Hence

$$AB \times DC = BC \times AD$$
$$DC = \frac{2 \times 15}{10} = \frac{30}{10}\,\Omega$$
$$= 3\,\Omega$$

o

Hence to balance the bridge a 1 Ω resistor must be added in series with the 2 Ω resistor in the arm *DC*.

When the bridge is balanced *AB* and *BC* are in series (same current) and *AD* is in series with *CD*. Hence the network resolves itself into a resistance of $(10 + 2)\Omega$ in parallel with one of $(15 + 3)\Omega$, and the total current will divide in the inverse ratio of the resistances. Therefore

$$\text{current in } ABC = 1{\cdot}5 \times \frac{18}{12 + 18}A = 0{\cdot}9\,A$$

$$\text{current in } ADC = 1{\cdot}5 \times \frac{12}{12 + 18}A = 0{\cdot}6\,A$$

Alternatively,

$$\text{current in } ADC = 1{\cdot}5 - 0{\cdot}9 = 0{\cdot}6\,A$$

Example 2.2

When a reduced voltage of 4 V was connected across *AC*, in the original network in Example 2.1, the current taken from the battery was 0·57 A. What then would be the current in the 7 Ω resistor?

Fig. 2.3 shows the network. Since the bridge is unbalanced it is necessary

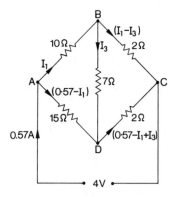

Fig. 2.3 Network for Example 2.2.

to apply Kirchhoff's laws and Fig. 2.3 shows the assumed current distribution, applying the First Law to the points *A*, *B* and *D*.

Applying the Second Law to network *ABC*,

$$4 = 10I_1 + 2(I_1 - I_3)$$
$$= 12I_1 - 2I_3$$

Hence
$$I_1 = \frac{4 + 2I_3}{12} = \frac{2 + I_3}{6}$$

Considering ADC,

$$4 = 15(0\cdot57 - I_1) + 2(0\cdot57 - I_1 + I_3)$$
$$= 8\cdot55 - 15I_1 + 1\cdot14 - 2I_1 + 2I_3$$
$$= 9\cdot69 - 17I_1 + 2I_3$$

Substitute for I_1.

$$4 = 9\cdot69 - 17\left[\frac{2 + I_3}{6}\right] + 2I_3$$
$$24 = 58\cdot14 - 34 - 17I_3 + 12I_3$$
$$-0\cdot14 = -5I_3$$
$$I_3 = 28 \text{ mA}$$

Current in the 7 Ω resistor is 28 mA in the direction shown in Fig. 2.3.

Example 2.3
On a metre wire bridge, used to determine the value of an unknown resistance, balance was obtained at a point 375 mm from the end of the wire opposite the known resistance of 10 Ω. What is the value of the unknown resistance?

With the metre wire bridge it is the ratio arms that vary but the condition for balance is the same as for Wheatstone's bridge. Since balance is obtained 375 mm from one end of the wire then the remaining length is $1000 - 375 = 625$ mm.

Assuming that the wire is of uniform cross-sectional area and resistivity then the resistance ratio of the two arms formed by the wire, is the same as the ratio of their lengths. Let X Ω be the value of the unknown resistance, then

$$625 \times X = 375 \times 10$$
$$X = \frac{375 \times 10}{625} = 6\,\Omega$$

The unknown resistance has a value of 6 Ω.

Example 2.4
In a Post Office box (three arrangements of resistors in one case) the ratio arms have resistances of 10, 100 and 1000 Ω and the variable resistance has a range of 1 to 500 Ω. What are the maximum and minimum values of resistance which can be measured by this apparatus?

Example 2.5
A square, $ABCD$, is formed by four resistors connected in series. Their values are; $AB = 4\,\Omega$, $BC = 2\,\Omega$, $CD = 3\,\Omega$ and $DA = R\,\Omega$. A 2 V battery, in series with a 1·4 Ω resistor, is connected between A and C and between

35

B and *D* is a 10 Ω resistor. Calculate the value of *R* and the current in it when the current in the 10 Ω resistor is zero.

Example 2.6
Fig. 2.4 shows the circuit used in locating the fault in a twin-core cable. *AB* and *CD* are two telephone wires, 15 km long and of resistance 50 Ω/km. The

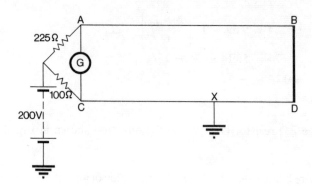

Fig. 2.4 Circuit for Example 2.6.

ends *B* and *D* are short circuited and balance is obtained with the values indicated in the diagram. Determine the distance of the fault, at *X*, from *C*.

Example 2.7
Four resistors, *AB*, *BC*, *CD* and *DA* are connected in series to form a closed network. Their respective values are 3 Ω, 4 Ω, 5 Ω and 5 Ω and a 20 Ω resistor is connected between *B* and *D*. If a 2 V battery, having an internal resistance of 0·5 Ω is connected across *A* and *C*, determine the current in the 20 Ω resistor.

Example 2.8
In a distance thermometer, the bulb consists of a nickel resistance of 10 Ω at 0°C, and forms the fourth arm *CD* of a Wheatstone's bridge. The other three arms are composed of manganin resistors having the following values: *AB* = 3 Ω, *BC* = 9 Ω and *AD* = 4 Ω, all values being measured at 0°C. If the temperature coefficient of resistance of nickel is 0·0062°C^{-1} and that of manganin is negligible, estimate the temperature at which the galvo, connected between *A* and *C*, will show balance.

2.2 The Potentiometer

The potentiometer provides a means of measuring potential difference, voltage or e.m.f.

In its simplest form it consists of a long wire (4 m) of uniform cross-section, which carries a constant current. Since

$$V = I \times \rho l/a \text{ volts}$$

then $\qquad V_1/V_2 = l_1/l_2$

Hence the voltage drop, between any two points in the wire, is proportional to the length of the wire between the two points. Fig. 2.5 shows a standard

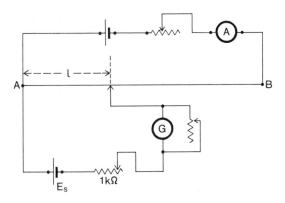

Fig. 2.5 Circuit for the calibration of a simple potentiometer.

cell, of known constant e.m.f. connected to the potentiometer. (Note the polarities of the standard cell and the driving battery.)

There must be some length along the potentiometer wire in which the voltage drop is equal to the e.m.f. of the standard cell (E_s) and this length is determined by adjusting the position of the sliding contact until no current flows in the galvo circuit.

Let the position of balance be such that the length along the wire be l mm from A, the positive end of the potentiometer and the end from which all measurements should be made. Hence the volt drop along the wire = E_s/l V/mm. This is the calibration constant of the potentiometer and once it has been determined the potentiometer may be used for other measurements. This calibration constant is only true for one particular value of current in the potentiometer wire, and so for any one set of readings this current must be checked frequently and maintained constant.

Note that it is preferable that no current be drawn from the standard cell and so it is usual to protect the cell by connecting a 1 kΩ resistor in series with it. This resistance may be removed to obtain the final balance point. In commercial types of potentiometer the resistance wire is usually arranged in sections and brought out to two dials which are set to equal the e.m.f. of the standard cell (e.g. for a cell having an e.m.f. of 1·0186 V the dials are set at 1·0 and 18·6 respectively). The coarse and fine resistors are then adjusted to give zero deflection of the galvanometer. The apparatus, so adjusted, then becomes a direct reading instrument, with 1 division $\equiv$ 0·001 V.

Example 2.9

On a 4 m potentiometer the e.m.f. of a standard cell, 1·0183 V, was balanced at a point 2·25 m from the positive end of the potentiometer. At what distance from that same end would balance be obtained if the standard cell were replaced by a dry cell of e.m.f. 1·5 V?

$$\text{Calibration constant for potentiometer} = \frac{1 \cdot 0183}{2 \cdot 25} \frac{V}{m}$$

Hence

$$\text{Distance of tapping point for 1·5 V} = \frac{1 \cdot 5}{\text{calibration constant}} m$$

$$= 1 \cdot 5 \times \frac{2 \cdot 25}{1 \cdot 0183} m = 3 \cdot 314 \, m$$

The distance of the tapping point from the positive end of the potentiometer will be 3·314 m.

Example 2.10

When checking the accuracy of a 0–1 A ammeter a current is passed through a known standard resistance and the volt drop across the resistance measured by a potentiometer. In such a test the volt drop across a 10 Ω standard resistance was balanced at a distance of 595 mm from the positive end of the potentiometer when the ammeter was reading 0·15 A. Determine the percentage error of the instrument if the calibration constant for the potentiometer is 0·0026 V/mm.

$$\text{Volt drop across resistor} = \text{tapping distance} \times \text{calibration constant}$$

$$= 595 \times 0 \cdot 0026 \, V = 1 \cdot 547 \, V$$

$$\text{True current} = \frac{1 \cdot 547}{10} A = 0 \cdot 1547 \, A$$

$$\text{Instrument error} = 0 \cdot 1547 - 0 \cdot 15 \, A = 0 \cdot 0047 \, A$$

$$\text{Error} = \frac{0 \cdot 0047}{1 \cdot 5} \times 100\% = 3 \cdot 13 \% \text{ (low)}$$

At a reading of 0·15 A, the instrument is reading 3·13 per cent low.

Example 2.11

A 4 m potentiometer has a resistance of 10 Ω and is connected to a 4 V battery. A 1·018 V standard cell, in series with a galvanometer of resistance

20 Ω is tapped onto the potentiometer. Determine the current flowing in the galvanometer when the sliding contact is 2 mm short of the balance point.

If the potentiometer were balanced no current would flow in that part of the circuit containing the standard cell. Hence the circuit may be considered as a 4 V battery connected to a 10 Ω resistance. Hence

$$\text{current in potentiometer wire} \quad = \frac{4}{10} = 0.4 \text{ A}$$

At balance,

volt drop along 4 m wire $\qquad = 4$ V

volt drop along wire to tapping point $= 1.018$ V

$$\text{distance to tapping point for balance} \ = \frac{4}{4} \times 1.018 \text{ m} = 1.018 \text{ m}$$

$$\text{resistance of 1 mm of wire} \qquad = \frac{10}{4000} = 2.5 \text{ m}\Omega.$$

When sliding contact is 2 mm short of balance point,

Length of wire tapped off $\qquad = 1.018 - 0.002$ m $= 1.016$ m

resistance of wire tapped off $\qquad = 2.5 \times 10^{-3} \times 1016 = 2.54 \ \Omega$

Fig. 2.6 shows the equivalent circuit for the unbalanced condition. Let I_1

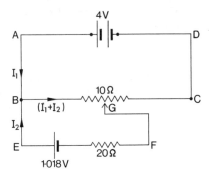

Fig. 2.6 Equivalent circuit for unbalanced potentiometer in Example 2.11.

be the current through the 4 V battery and I_2 the current through the standard cell.

Applying Kirchhoff's First Law the current distribution will be as shown. Apply the Second Law to the network *ABCD*.

$$4 = 2.54(I_1 + I_2) + 7.46I_1$$
$$= 10I_1 + 2.54I_2 \tag{1}$$

Consider network *BEFG*.

$$1{\cdot}018 = 2{\cdot}54(I_1 + I_2) + 20I_2$$
$$= 2{\cdot}54I_1 + 22{\cdot}54I_2 \tag{2}$$

Multiply (1) by 2·54 and (2) by 10.

$$10{\cdot}16 = 25{\cdot}4I_1 + 6{\cdot}45I_2$$
$$10{\cdot}18 = 25{\cdot}4I_1 + 225{\cdot}4I_2$$

Subtracting, $\qquad -0{\cdot}02 = -218{\cdot}95I_2$

$$I_2 = 91{\cdot}3\ \mu A$$

The current through the galvo would be 91·3 μA in the direction shown in Fig. 2.6.

Example 2.12
A 4 m potentiometer has a resistance of 10 Ω. With a standard cell of e.m.f. 1·0186 V balance is obtained at a length of 509·3 mm from the positive end of the potentiometer. What is the maximum voltage which may be measured on the potentiometer and what must be the current in the wire at balance?

Example 2.13
The 1·018 V e.m.f. of a standard cell is balanced by the volt drop along 0·1527 m of a potentiometer wire. With the same current in the potentiometer wire, the volt drop across a 5 Ω resistor is balanced by the volt drop in 0·25 m of the wire. Estimate the current in the resistor.

Example 2.14
A 2·5 V voltmeter is tested against a potentiometer, which has been calibrated by a standard cell of e.m.f. 1·018 V, and the balance point is obtained at 509 mm from the positive end. With the voltmeter reading 2·5 V the new balance was obtained at 1269 mm from the same end. Determine the percentage error of the instrument at full-scale deflection. At what distance would you expect to obtain balance when the voltmeter is reading 1·5 V?

Example 2.15
A current is passed through an unknown resistance which is connected in series with a standard resistance of 10 Ω. The volt drop across the standard resistance was balanced on a potentiometer at a distance of 1452 mm from the positive connection and the volt drop across the unknown resistance was balanced at a distance of 1726 mm. Estimate the value of the unknown resistance.

Example 2.16
A potentiometer wire of length 4 m and having a resistance of 12 Ω is supplied

by a 4 V battery. A standard cell, of e.m.f. 1·0186 V, in series with a galvano-meter, of 25 Ω resistance, is tapped onto the potentiometer wire and balance obtained. If the tapping distance is then increased, by an amount equivalent to 0·01 Ω of the potentiometer wire, determine the magnitude and direction of the current in the galvanometer circuit.

Example 2.17
The e.m.f. of a standard cell, 1·0183 V, is balanced by the volt drop in 762 mm of a potentiometer wire. When a current of 0·3 A, as measured by a sub-standard ammeter, is passed through a resistor the volt drop across the resistor is balanced by that in 2·848 m of the potentiometer wire. Estimate the resistance value of the resistor.

2.3 The Potential Divider

The potential divider is often referred to as a potentiometer or simply as a 'pot'. It consists, normally, of a high variable resistor connected directly across the supply and the varying voltages taken off between one end of the divider and either, one of a number of tappings brought out to studs, or the 'slider'. The latter type is used as a volume control in a radio receiver, etc.

Example 2.18
A 1 kΩ divider is connected across a 250 V d.c. supply. Determine the position of the tapping point, such that a 25 Ω resistor, connected across the tapping point and the supply, carries a current of 2 A.

Fig. 2.7 shows the circuit with AB as the potential divider. Let R Ω be the

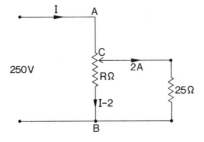

Fig. 2.7 Circuit for the potential divider in Example 2.18.

resistance between the tapping point C and B. Then $(1000 - R)\Omega$ is the resistance between A and C. Let I be the current taken from the supply. Then the current in CB will be $(I - 2)$.

Volt drop across resistor $= 2 \times 25 = 50$ V

Hence

volt drop across CB $= 50\,\text{V} = R(I-2)$

Volt drop across AC $= 250 - 50 = 200\,\text{V}$

$= I(1000 - R)$

$$I = \frac{200}{(1000 - R)}$$

Substituting in $R(I-2) = 50$,

$$50 = R\left(\frac{200}{1000 - R} - 2\right)$$

$$50(1000 - R) = 200R - 2000R + 2R^2$$

$$50\,000 + 1750R - 2R^2 = 0$$

$$25\,000 + 875R - R^2 = 0$$

Solving this quadratic equation gives,

$$R = \frac{875 \pm \sqrt{\{875^2 - (4 \times -1 \times 25\,000)\}}}{2}$$

$$= \frac{875 + 930}{2} = 902\cdot5\,\Omega$$

The second solution is negative and inadmissible.

The tapping point is at a point $902\cdot5\,\Omega$ from B.

Example 2.19

A circuit of resistance $0\cdot5\,\Omega$ is to be supplied with $10\,\text{A}$ from a $30\,\text{V}$ d.c. supply through a $10\,\Omega$ potential divider. Determine the value of the resistance to be tapped off on the divider.

Example 2.20

A potential divider, having a resistance of $90\,\Omega$ is connected directly across a $15\,\text{V}$ d.c. supply. A coil is connected between the positive end of the divider and a point which is $60\,\Omega$ away from that same end. If the coil takes a current of $0\cdot1\,\text{A}$ determine the resistance of the coil.

Example 2.21

It is necessary to supply a load resistance with $0\cdot5\,\text{A}$ at $100\,\text{V}$ from a $240\,\text{V}$ supply. Calculate the required tapping position on a $5\,\text{k}\Omega$ potential divider.

Chapter 3

Electromagnetism

Whenever movement of electrons is produced by a directive force, i.e. a p.d., a magnetic field is always associated with that movement. The magnitude and direction of this field depends on the rate of flow of the electrons, i.e. the current (1 A = 6·2 × 10^{18} electrons/s) and the direction of such a flow. The shape of the field depends on the path taken by the electrons. In general, if the electrons travel in a linear path, the magnetic field will be circular in cross-section and if the electrons travel in a circular path the magnetic field through the path will be linear.

3.1 Field due to Electrons Moving in a Straight Line

The magnetic field is of cylindrical form, centred on the centre of the current path. This path will normally be through a conductor but need not necessarily be so. The direction of the field is given by the Right-hand Corkscrew Rule.

When a current of I A flows, the *flux density* (B), at a distance of d m from the centre of the current path, is given by

$$B = \frac{\mu_0 I}{2\pi d} \text{ tesla (T)}$$

where μ_0 is the permeability of free space and is a constant equal to $0·4\pi \times 10^{-6}$ SI units, or $1·257 \times 10^{-6}$ SI units.

The flux density is equal to the total flux (Φ webers (Wb)) divided by the area of the field A m^2.

$$B = \Phi/A$$

Hence, 1 tesla = 1 Wb/m^2.

Example 3.1
What is the flux density at a distance of 100 mm from the centre of a conductor, carrying a current of 20 A?

$$B = \frac{\mu_0 I}{2\pi d} = \frac{0.4\pi \times 10^{-6} \times 20}{2\pi \times 100 \times 10^{-3}} \text{ T}$$

$$= 0.04 \text{ mT}$$

(Note: Theoretically all distances should be measured from the centre of the conductor, but in practice the radius of the conductor will be negligible in comparison to the distance from the conductor being considered. Hence only a negligible error will be introduced if the distance is measured from the circumference of the conductor.)

Example 3.2
Determine the magnitude of the current in a conductor, required to produce a flux density of 0·1 mT at a distance of 50 mm from the conductor.

$$B = \mu_0 I / 2\pi d$$

Hence

$$I = 2\pi B d / \mu_0$$

$$= \frac{2\pi \times 0.1 \times 10^{-3} \times 50 \times 10^{-3}}{0.4\pi \times 10^{-6}}$$

$$= 25 \text{ A}$$

Example 3.3
Calculate the magnitude of the flux density at a distance of 75 mm from the centre of a long straight conductor carrying a current of 15 A.

Example 3.4
A straight conductor carries a current of 25 A. Plot a curve showing how the flux density varies from the surface of the conductor to a point 50 mm from the surface.

Example 3.5
At what distance from a straight conductor, carrying a current of 10 A, will the flux density be 50 μT?

Example 3.6
The flux density, at a point 100 mm, from a straight conductor, is found to be 0·06 mT. Estimate the value of current in the conductor.

Example 3.7
A conductor, carrying a current of 25 A, is moved, at a constant distance of 100 mm, around an unscreened moving-coil instrument. If the instrument has been calibrated on the assumption of a constant flux density of 30 mT in the air gap, estimate the maximum possible percentage error that may occur

on full-scale deflection. (The deflection is proportional to the flux density in the gap for any one value of current in the moving coil.)

3.2 Force Acting on Electrons which are Being Moved Perpendicularly to a Magnetic Field

When electrons move at right angles to a constant magnetic field, the magnetic field produced by the movement of the electrons distorts the main field, which is then in a state of tension. Consequently the main field, in trying to return to a state of stability, will exert a force on the electrons. If the electron path is through a conductor then this force will be transferred to the conductor.

This force will act in a direction which is perpendicular to both the direction of electron flow and the magnetic field. The direction of the force is given by Fleming's Left-hand Rule. The magnitude of the force (F) varies directly with the flux density (B) of the field, the rate of flow of the electrons (i.e. the current I) and the length of the electron path in the field (l).

If these quantities are measured in their fundamental units, teslas, amperes and metres, then

$$F = BIl \text{ newtons (N)}$$

Example 3.8
A conductor, of length 200 mm, is placed at right angles to a magnetic field of density 0·2 T. If a current of 10 A is maintained in the conductor, calculate the force which will act on the conductor.

$$F = BIl = 0\cdot2 \times 10 \times 200 \times 10^{-3} \text{ N}$$
$$= 0\cdot4 \text{ N}$$

Example 3.9
A conductor carries a current of 20 A. A second conductor, with a current of 15 A, is placed parallel to the first conductor and with 10 mm between their centres. Determine the force/metre acting on the conductors and by means of diagrams determine the direction of the force if the two currents are (a) in the same direction, (b) in opposite directions.

Under the conditions stated in the question, the second conductor will be perpendicular to the field produced by the first conductor. The density of this magnetic field will be given by

$$B = \frac{0\cdot2 \times 10^{-6}I}{d} \text{ T}$$
$$= \frac{0\cdot2 \times 10^{-6} \times 20}{10 \times 10^{-3}} \times 10^3 \text{ mT}$$
$$= 0\cdot4 \text{ mT}$$

45

Hence the force/m on the second conductor will be

$$0.4 \times 10^{-3} \times 15 \times 1\,\text{N/m} = 6\,\text{mN/m}$$

Although this force has been calculated as acting on the second conductor it is a mutual force and therefore the force/m acting on the first conductor will also be 6 mN/m.

Note: If, $B = 0.2 \times 10^{-6} I_1/d$ is substituted in $F = BI_2l$ then

$$F = \frac{0.2 \times 10^{-6} \times I_1 \times I_2 \times l}{d}\text{N}$$

Since the force is proportional to the product of the two currents, its value is unchanged no matter which of the two currents is considered as producing the field.

Figs. 3.1(a) and (b) represent the conditions required. The direction of the fields is determined by using the Right-hand Corkscrew Rule.

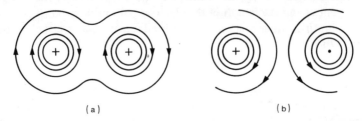

(a)　　　　　　　　　　　　　　(b)

Fig. 3.1 Magnetic fields due to current carrying conductors—Example 3.9.

In Fig. 3.1(a) the fields in the inter-conductor space are in opposition and therefore, at some point in this space, there will be a point at which the resultant field is zero. Hence the field will be forced to link the conductors as shown. The field will be in tension, and in trying to return to a state of stability will exert a force of attraction on the conductors.

Fig. 3.1(b) shows that in the inter-conductor space the two fields are additive and in consequence the field will be in compression and so will produce a force of repulsion on the conductors.

Example 3.10
Two underground single-core cables, which are supplied at 240 V from a source having an internal resistance of 0.05 Ω, are placed with 100 mm between their centres. If the resistance of each conductor is 50 μΩ/m, calculate the force between the two conductors when a short circuit fault occurs at a distance of 100 m from the source.

Under fault conditions,

$$\text{total resistance of the circuit} = 0.05 + 2 \times 50 \times 100 \times 10^{-6}\,\Omega$$
$$= 0.05 + 0.01\,\Omega = 0.06\,\Omega$$

and so

$$\text{short circuit current} \quad = \frac{240}{0 \cdot 06} \text{A} = 4000 \text{ A}$$

As the two conductors are in series they will carry the same current. Hence

$$F = \frac{0 \cdot 2 \times 10^{-6} \times I^2 \times l}{d}$$

$$= \frac{0 \cdot 2 \times 10^{-6} \times 16 \times 10^6 \times 100}{100 \times 10^{-3}} \text{N} = 3 \cdot 2 \text{ kN}$$

Example 3.11
A conductor of length 0·8 m, carrying a current of 15 A, is placed at an angle of $\frac{1}{4}\pi$ rad to a field of effective length 1 m and density 0·8 T. Calculate the force acting on the conductor.

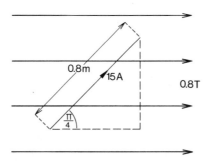

Fig. 3.2 Current carrying conductor lying at an angle to a magnetic field—Example 3.11.

Fig. 3.2 shows the relative position of the conductor to the field. The effective length of the conductor perpendicular to the field will be

$$0 \cdot 8 \sin \tfrac{1}{4}\pi \text{ m} = 0 \cdot 8 \times 0 \cdot 707 \text{ m} = 0 \cdot 566 \text{ m}$$

The force on the conductor will be

$$0 \cdot 8 \times 15 \times 0 \cdot 566 \text{ N} = 6 \cdot 792 \text{ N} = 6 \cdot 8 \text{ N}$$

Example 3.12
A conductor, of length 250 mm, lies at right angles to a magnetic field of density 0·75 T. If a force of 1·65 N is exerted on the conductor when it carries a current of 10 A, determine the length of the conductor in the magnetic field.

Example 3.13
When a conductor in an electric motor is under the influence of a pole, it

is required that the force exerted on the conductor should be 10 N when the current in the conductor is 40 A. If the axial length of the pole is 0·3 m, estimate the required flux density under the pole.

Example 3.14
When a circuit, carrying 800 A, is broken a spark is produced across the gap between the contacts. If the magnetic blow-out coils produce a field of density 0·1 T at right angles to the spark, calculate the force exerted on each mm length of the spark.

Example 3.15
In a cathode ray tube a stream of electrons, in passing from the emitter to the screen, passes at right angles across the field due to a pair of coils mounted on the neck of the tube. This field has a density of 5 mT and a magnetic length of 25 mm. If the resultant force on the electron stream is 0·2 μN, estimate the rate of flow of electrons through the field.

Example 3.16
Two copper bus-bars, having a circular cross-section and of length 0·6 m, are bolted at each of their ends to a distribution board, with 80 mm between their centres. If the four bolts used each have a diameter of 6 mm, determine the shear stress in the bolts when the bus-bars carry a current of 300 A.

Example 3.17
Two parallel conductors, A and B, with 75 mm between their centres, carry currents of 30 A and 20 A respectively. If the currents are in the same direction, calculate the distance of the neutral point from A. A third parallel conductor, carrying a current of 50 A but in the opposite direction to that in A, is placed at this neutral point. Calculate the total force/m acting on this third conductor.

Example 3.18
When two parallel conductors carry the same current the mutual force between them is 1·6 mN. If the length of the conductors is 0·4 m and their centres are 30 mm apart, calculate the current which the conductors are carrying. What would be the force between the conductors if one conductor were at an angle of $\pi/6$ rad to the field of the other?

3.3 Torque on a Rectangular Coil in a Magnetic Field

When a current is maintained in a coil, the plane of which lies parallel to a uniform magnetic field, then a force acts on each of the two sides of the coil, which are perpendicular to the field. These forces form a couple and so exert a torque on the coil.

48

Example 3.19
A coil of 20 turns, which measures 30 mm × 20 mm, is supported, about the mid-points of its shorter sides, in a radial field of density 0·5 T. Estimate the torque produced when the coil carries a current of 10 A.

Force acting on one side of one turn $= 0·5 \times 10 \times 30 \times 10^{-3}$ N $= 0·15$ N

Hence total force on one coil side $\quad = 20 \times 0·15$ N $= 3$ N

torque on one coil side $\quad = $ force × distance from point of rotation

$$= 3 \times \frac{20}{2} \times 10^{-3} \text{ Nm}$$

$$= 30 \times 10^{-3} \text{ Nm}$$

total torque on coil (2 sides) $\quad = 2 \times 30 \times 10^{-3}$ Nm

$$= 60 \times 10^{-3} \text{ Nm}$$

Note: A coil mounted in a magnetic field will always try to rotate so that its plane becomes perpendicular to the field, i.e. the coil embraces the maximum quantity of flux.

Example 3.20
A coil, having an area of 800 mm², is wound with 500 turns and is mounted in a field of density 0·5 T. Calculate, from first principles, the torque produced when the coil carries a current of 20 mA.

The force on a current carrying conductor in a magnetic field is BIl, where B is the flux density in teslas, I is the current in amperes, l is the length in metres.

Hence for a coil of T turns and width w,

Force on coil side $\quad = TBIl$

Torque per coil side $= TBIl \times \dfrac{w}{2}$

Total torque on coil $= 2 \times TBIl \times \dfrac{w}{2}$

$$= TBIlw$$

But $l \times w = $ area of coil A. Hence

total torque on coil $= TBIA$

Also $B \times A = $ total flux $= \Phi$, and so

total torque on coil $= T\Phi I$

$$= 500 \times 0·5 \times 800 \times 10^{-6} \times 20 \times 10^{-3} \text{ Nm}$$

$$= 4 \times 10^{-3} \text{ Nm}$$

P

Example 3.21

A coil on the rotor of a generator has 25 turns, an axial length of 200 mm and a mean width of 120 mm. If the flux density in the air gap is 0·8 T, calculate the power required to rotate the coil at an angular velocity of 104 rad/s, when the coil carries a current of 2·5 A.

Since the torque on a coil = $T\Phi I$, then

$$\text{Torque to be overcome} = 25 \times 0.8 \times 200 \times 10^{-3}$$
$$\times 120 \times 10^{-3} \times 2.5$$
$$= 1.2 \text{ Nm}$$

Work done in turning coil through 1 rad = 1·2 J

Work done/s $= 1.2 \times 104 \text{ J/s}$

Power required $= 125 \text{ W}$

Note: Power = torque × angular velocity.

Example 3.22

A coil, wound with 20 turns, is 20 mm long and 15 mm wide. Calculate the torque produced when this coil, carrying a current of 15 mA is placed in a magnetic field of density 0·3 T.

Example 3.23

A rectangular coil, wound with 50 turns, measures 25 mm × 15 mm and is mounted in a uniform magnetic field. When the coil carries a current of 0·5 A the torque produced is just balanced by a control torque of 50 μNm. Estimate the flux density of the field.

Example 3.24

The area of a coil of 20 turns is 6×10^{-3} m². The coil is mounted in a field of density 0·4 T and is held with its plane parallel to the field by an opposing torque of 0·05 Nm. Determine the value of the current in the coil.

Example 3.25

A coil on a motor armature consists of 15 turns and its mean dimensions are 100 mm × 60 mm. If the armature carries 20 such coils and they each carry a full-load current of 5 A, estimate the flux density in the air gap when the output of the motor is 1·25 kW at an angular velocity of 150 rad/s.

Example 3.26

The flux density in the air gap of a d.c. generator is 0·5 T. The armature, which carries 15 effective coils each of 15 turns, rotates at 74 rad/s when the mechanical input to the armature is 4 kW. Determine the load current and the output voltage of the generator. The area of each coil is 200 × 150 mm².

Example 3.27
A d.c. motor takes a full-load current of 16 A from a 240 V d.c. supply. The armature circuit consists of 10 effective coils (250 mm × 200 mm), each of 8 turns. If the flux density in the air gap is 0·7 T, determine the angular velocity with which the motor will run.

3.4 The Field due to a Solenoid

The field due to a solenoid is similar in shape to that of a bar magnet, and as long as the length of the solenoid is large in comparison to its diameter it is assumed that, over the length of the solenoid, the internal field is uniform.

It is also assumed that the length of the flux path is equal to the length of the solenoid only. This assumption is based on the fact that outside the solenoid the flux is distributed over such a large area that the flux density is very low.

Example 3.28
An air-cored coil, wound with 500 turns, has a length of 250 mm and a cross-sectional area of 400 mm². Calculate the magneto-motive force, the magnetizing force and the total flux produced when a current of 5 A is maintained in the coil.

The magneto-motive force, F, is equal to the product of the current and the number of turns. Since the number of turns has no unit, the unit of this force is the Ampere. Hence

$$F = 5 \times 500 \text{ A} = 2500 \text{ A}$$

The magnetizing force, H, is equal to the magneto-motive force per unit length and hence its unit is the Ampere/metre (A/m). Therefore

$$H = \frac{2500}{250 \times 10^{-3}}$$
$$= 10^4 \text{ A/m} = 10 \text{ kA/m}$$

Total flux $\Phi = B \times A$

But $B = \mu_0 H$, and so

$$\Phi = \mu_0 H A$$
$$= 0 \cdot 4\pi \times 10^{-6} \times 10 \times 10^3 \times 400 \times 10^{-6} \text{ Wb}$$
$$= 5 \, \mu\text{Wb}$$

Example 3.29
Calculate the value of the current required to produce a flux of 6 μWb in an air cored coil, 400 mm long and 250 mm² in cross-sectional area, when the coil is wound with 750 turns.

$$\Phi = \mu_0 H A = \frac{\mu_0 I T A}{l}$$

Hence
$$I = \frac{\Phi l}{\mu_0 T A}$$

$$= \frac{6 \times 10^{-6} \times 400 \times 10^{-3}}{0 \cdot 4\pi \times 10^{-6} \times 750 \times 250 \times 10^{-6}} \text{ A}$$

$$= 10 \cdot 2 \text{ mA}$$

Example 3.30
A coil of 200 turns is wound on an iron core of length 250 mm and cross-sectional area 400 mm². If the iron has a relative permeability of 800, estimate the total flux produced when a current of 1·5 A is maintained in the coil.

$$\text{Relative permeability} = \frac{\text{flux density with iron core}}{\text{flux density with air core}}$$

but flux density with air core $B = \mu_0 H$, and therefore

$$\mu_r = B/\mu_0 H$$
$$B = \mu_0 \mu_r H$$
$$BA = \frac{\mu_0 \mu_r I T A}{l}$$

$$= \frac{1 \cdot 257 \times 10^{-6} \times 800 \times 1 \cdot 5 \times 200 \times 400 \times 10^{-6}}{250 \times 10^{-3}} \text{ Wb}$$

$$= 0 \cdot 48 \text{ mWb}$$

Note: As the relative permeability of a material is the ratio between two flux densities, it has no unit.

Relative permeability is not a constant. For any one material it varies with the working flux density and the working temperature. It varies with different materials, and all non-magnetic materials are assumed to have a relative permeability of unity.

The ratio between the flux density and the magnetizing force is called the *absolute permeability* of the material ($\mu_0 \times \mu_r$).

Example 3.31
When a magneto-motive force of 10^3 A is applied to an iron-cored coil of length 200 mm and diameter 50 mm, the measured flux is 12 mWb. Calculate the reluctance of the coil and hence estimate the relative permeability of the iron core.

$$\text{Reluctance} = \frac{\text{magneto-motive force}}{\text{total flux}}$$

$$= \frac{10^3}{12 \times 10^{-3}} = 83 \cdot 3 \times 10^3 \text{ SI units}$$

Since

$$\Phi = \frac{\mu_0 \mu_r I T A}{l}$$

then

$$\frac{IT}{\Phi} = \frac{l}{\mu_0 \mu_r A}$$

Hence

$$\text{reluctance} = \frac{l}{\mu_0 \mu_r A}$$

$$\mu_r = \frac{l}{\mu_0 SA} \qquad \text{where } S \equiv \text{reluctance}$$

$$= \frac{200 \times 10^{-3}}{1 \cdot 257 \times 10^{-6} \times 83 \cdot 3 \times 10^3 \times \pi \times (2500/4) \times 10^6}$$

$$= \frac{8}{1 \cdot 257 \times 83 \cdot 3 \times \pi \times 25 \times 10^{-6}} = 970$$

The relative permeability of the iron is 970.

Example 3.32
What value of current must be maintained in a coil of 250 turns to produce a magneto-motive force of 600 A?

Example 3.33
A current of 4 A is maintained in a coil of 250 turns which are wound uniformly on a non-magnetic former of length 200 mm. What is the magnetizing force and what will be the value of the flux density produced?

Example 3.34
A solenoid has a length of 450 mm and a cross-sectional area of 600 mm². Calculate the reluctance of the solenoid and the magneto-motive force required to produce a total flux of 0·24 mWb.

Example 3.35
How many ampere turns are required to produce a total flux of 10 mWb in the pole of a machine which has a magnetic length of 200 mm and a cross-sectional area of 1200 mm²? At the working flux density the relative permeability of the magnetic material is 1500.

Example 3.36
A magnetizing force of 1000 A/m is applied to a coil of wire wound on an iron former of cross-sectional area 318 mm². If the iron, at the flux density produced, has a relative permeability of 1200, estimate the total flux produced in the iron.

Chapter 4

The Magnetic Circuit

In Chapter 3, when dealing with the magnetic field due to a solenoid, the flux path through air was neglected, but in magnetic circuits all parts of the flux path must be considered and especially any air gaps in the circuit, as these usually require the greater part of the magneto-motive force (m.m.f.) in order to maintain the flux in the gap.

4.1 The Magnetic Circuit

Magnetic circuits may be sub-divided into the same three types as are the electric circuits: series, parallel and series-parallel.

Example 4.1

A magnetic circuit consists of a magnetic path of mean length 1 m and a 2 mm air gap. The circuit has a uniform cross-sectional area of 10^{-3} m^2. Calculate the magnetizing current required to produce a total flux of 1 mWb in the air gap, if the iron is uniformly wound with 500 turns, and at the working flux density has a relative permeability of 880.

In order to determine the magnetizing current it is first necessary to calculate the total m.m.f. In a series electric circuit the total m.m.f. may be determined by multiplying the total resistance of the circuit by the current in the circuit. In a series circuit the total resistance is equal to the sum of the separate resistances. For the magnetic circuit,

$$\text{total m.m.f.} = \text{total flux} \times \text{total reluctance}$$

$$= \Phi \,(\text{reluctance of iron} + \text{reluctance of gap})$$

$$\text{reluctance of iron} = \frac{l}{\mu_0 \mu_r A} = \frac{1}{1\cdot257 \times 10^{-6} \times 880 \times 10^{-3}}$$

$$= 905 \times 10^3 \text{ SI units}$$

$$\text{reluctance of air gap} = \frac{2 \times 10^{-3}}{1\cdot257 \times 10^{-6} \times 1 \times 10^{-3}}$$

$$= 1 \cdot 59 \times 10^6 \text{ SI units}$$

total m.m.f. $\qquad = 1 \times 10^{-3} (905 \times 10^3 + 1 \cdot 59 \times 10^6) \text{ A}$

$$= 10^{-3} \times 2 \cdot 495 \times 10^6 \text{ A} = 2495 \text{ A}$$

But $\qquad\qquad\qquad\qquad \text{m.m.f.} = IT$

Hence $\qquad\qquad\qquad I = 2495/500 \text{ A} = 4 \cdot 99 \text{ A}$

The required magnetizing current is 5 A.

Alternative solution. Kirchhoff's laws may be applied to the magnetic circuits just as they are to the electric circuits and in terms of the magnetic circuit the second law may be expressed as: The total m.m.f. applied to a magnetic circuit is equal to the sum of the separate m.m.f.s required for each part of the circuit.

since $\qquad\qquad\qquad \Phi = \mu_0 \mu_r I T A / l$

and $\qquad\qquad\qquad B = \mu_0 \mu_r I T / l$

then $\qquad\qquad\qquad I T = B l / \mu_0 \mu_r$

This is the ampere–turn formula and by considering each part with respect to the flux density in the part, its length and its relative permeability, the m.m.f. for each part may be calculated.

As the circuit is of uniform cross-sectional area and the two paths are in series the flux density will be the same for both parts. Hence

$$\text{flux density for the circuit} = \frac{1 \times 10^{-3}}{10^{-3}} \text{T} = 1 \text{ T}$$

Applying the ampere–turn formula to the magnetic path,

$$IT = \frac{1 \times 1}{1 \cdot 257 \times 10^{-6} \times 880} \text{A} = 905 \text{ A}$$

For air the gap

$$IT = \frac{1 \times 2 \times 10^{-3}}{1 \cdot 257 \times 10^{-6} \times 1} \text{A} = 1590 \text{ A}$$

Hence $\quad$ total m.m.f. $\qquad = 905 + 1590 \text{ A} = 2495 \text{ A}$

$$\text{magnetizing current} = \frac{2495}{500} \text{A} = 4 \cdot 99 \text{ A}$$

The required magnetizing current is 5 A.

Example 4.2

Fig. 4.1 shows a magnetic circuit, in which, in the outer limbs, are effective air gaps of length 2 mm and 1·5 mm respectively. If the magnetic core is of square section and the sides of the outer limbs are 17·8 mm determine the m.m.f. required to produce a flux of 0·3 mWb in the shorter air gap. Neglect the effect of the iron. With this m.m.f. applied to the centre limb what will be the flux in the second air gap, and what must be the minimum cross-sectional area of the centre limb so that the flux density in it shall not exceed 1 T?

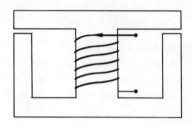

Fig. 4.1 Magnetic circuit for Example 4.2.

Just as in the electric circuit a true parallel circuit may only be obtained by neglecting the resistance of the leads, so in the magnetic circuit the parallel combination is only obtained by neglecting the reluctance of the magnetic paths. Hence, neglecting the reluctance of the iron, the two air gaps are in parallel.

Area of the flux path $= 17\cdot8^2 \times 10^{-6}\,\mathrm{m}^2$

Reluctance of shorter air gap $= \dfrac{l}{\mu_0\,\mu_r\,A} = \dfrac{1\cdot5 \times 10^{-3}}{1\cdot257 \times 10^{-6} \times 1 \times 17\cdot8^2 \times 10^{-6}}$

$= 3\cdot77 \times 10^6\ \mathrm{SI\ units}$

Reluctance of longer air gap $= \dfrac{2\cdot0 \times 10^{-3}}{1\cdot257 \times 10^{-6} \times 1 \times 17\cdot8^2 \times 10^{-6}}$

$= 3\cdot77 \times 10^6 \times \dfrac{4}{3} = 5\cdot03 \times 10^6\ \mathrm{SI\ units}$

Since the two reluctances are in parallel the m.m.f. required to produce 0·3 mWb in the shorter gap will also produce flux in the longer gap. Therefore

required m.m.f. F $=$ flux $\times$ reluctance

$= 0\cdot3 \times 10^{-3} \times 3\cdot77 \times 10^6\,\mathrm{A} = 1130\,\mathrm{A}$

flux in longer gap Φ_2 $= \dfrac{\mathrm{m.m.f.}}{\mathrm{reluctance}}$

$$\Phi_2 = \frac{1130}{5\cdot03 \times 10^6}\,\mathrm{Wb} = 0\cdot225\,\mathrm{mWb}$$

total flux $= 0\cdot3 + 0\cdot225\,\mathrm{mWb} = 0\cdot525\,\mathrm{mWb}$

$$\text{minimum area of centre limb} = \frac{\text{total flux}}{\text{maximum flux density}}$$

$$= \frac{0 \cdot 525 \times 10^{-3}}{1} \text{ m}^2 = 525 \text{ mm}^2$$

The flux in the longer air gap is 0·225 mWb and the minimum cross-sectional area of the centre limb is 525 mm².

Solution by the ampere–turn formula. Since area of the outer limbs is 17·8² mm², then

$$\text{flux density in shorter gap} = \frac{0 \cdot 3 \times 10^{-3}}{17 \cdot 8^2 \times 10^{-6}} \text{ T} = 0 \cdot 945 \text{ T}$$

But

$$AT = Bl/\mu_0\mu_r$$

$$\text{m.m.f. for shorter gap} = \frac{0 \cdot 945 \times 1 \cdot 5 \times 10^{-3}}{1 \cdot 257 \times 10^{-6} \times 1} \text{ A} = 1130 \text{ A}$$

Since 2 mm air gap is in parallel with the 1·5 mm gap, then

$$\text{flux density in 2 mm gap} = \frac{1130 \times 1 \cdot 257 \times 10^{-6}}{2 \times 10^{-3}} \text{ T} \qquad \left[\frac{\mu_0 IT}{l.} \right]$$

$$= 0 \cdot 709 \text{ T}$$

total flux in 2 mm gap $= 0 \cdot 71 \times 17 \cdot 8^2 \times 10^{-6} \text{ Wb} = 0 \cdot 225 \text{ mWb}$

total flux in centre limb $= 0 \cdot 3 + 0 \cdot 225 \text{ mWb} = 0 \cdot 525 \text{ mWb}$

$$\text{minimum cross-sectional area of centre limb} = \frac{0 \cdot 525 \times 10^{-3}}{1} \text{ m}^2 = 525 \text{ mm}^2$$

The flux in the 2 mm air gap is 0·225 mWb and the minimum cross-sectional area of the centre limb is 525 mm².

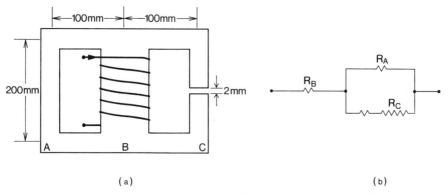

(a)   (b)

Fig. 4.2 (a) Magnetic circuit, with dimensions, for Example 4.3.
(b) Comparable electric circuit.

Example 4.3

Fig. 4.2(a) shows a magnetic circuit with its dimensions. Estimate the flux density in each part of the circuit when an m.m.f. of 300 A is applied on the centre limb. The centre limb has an area of $1 \cdot 2 \times 10^{-3}$ m^2 and the area of the remainder of the circuit is 600 mm^2. Assume a constant value of relative permeability for the magnetic material of 2000.

Fig. 4.2(b) shows the equivalent electrical circuit. Let S_t be the total reluctance of the circuit. Then

$$S_t = S_B + \frac{S_A \times S_C}{S_A + S_C}$$

Consider branch A.

Mean length of magnetic path $= 100 + 200 + 100$ mm $= 400$ mm

Then
$$S_A = \frac{400 \times 10^{-3}}{1 \cdot 257 \times 10^{-6} \times 2000 \times 600 \times 10^{-6}}$$
$$= 266 \times 10^3 \text{ SI units}$$

Consider branch B.

Mean length of magnetic path $= 200$ mm

Hence
$$S_B = \frac{200 \times 10^{-3}}{1 \cdot 257 \times 10^{-6} \times 2000 \times 1 \cdot 2 \times 10^{-3}}$$
$$= 66 \cdot 5 \times 10^3 \text{ SI units}$$

Consider branch C. In branch C there are two reluctances in series.

$$\text{Reluctance of air gap} = \frac{2 \times 10^{-3}}{1 \cdot 257 \times 10^{-6} \times 1 \times 600 \times 10^{-6}}$$
$$= 2 \cdot 66 \times 10^6 \text{ SI units}$$

Neglecting the reduction of 2 mm in the length of the magnetic path then the reluctance of the magnetic path in C is equal to the reluctance of A. Therefore

$$\text{total reluctance of } C \quad = 2 \cdot 66 \times 10^6 + 266 \times 10^3$$
$$= 2 \cdot 926 \times 10^6 \text{ SI units}$$

$$\text{total reluctance of circuit} = 66 \cdot 5 \times 10^3 + \frac{266 \times 10^3 \times 2 \cdot 926 \times 10^6}{(266 \times 10^3) + (2 \cdot 926 \times 10^6)}$$
$$= 66 \cdot 5 \times 10^3 + \frac{780 \times 10^9}{3 \cdot 192 \times 10^6}$$
$$= 66 \cdot 5 \times 10^3 + 244 \times 10^3$$
$$= 310 \times 10^3 \text{ SI units}$$

$$\text{But} \quad \text{total flux} = \frac{\text{total m.m.f.}}{\text{total reluctance}}$$

$$= \frac{300}{310 \times 10^3} \text{ Wb} = 0.97 \text{ mWb}$$

flux density in centre limb $= \dfrac{0.97 \times 10^{-3}}{1.2 \times 10^{-3}} \text{ T} = 0.808 \text{ T}$

For the flux in the outer limbs then the total flux will divide in the inverse ratios of the two reluctances.

Flux in A $= 0.97 \times \dfrac{2.926 \times 10^6}{3.19 \times 10^6} \text{ mWb} = 0.89 \text{ mWb}$

Flux in C $= 0.97 - 0.89 \text{ mWb} = 0.08 \text{ mWb}$

Flux density in $A = \dfrac{0.89 \times 10^{-3}}{600 \times 10^{-6}} \text{ T} = 1.48 \text{ T}$

Flux density in $C = \dfrac{0.08 \times 10^{-3}}{600 \times 10^{-6}} \text{ T} = 0.133 \text{ T}$

The flux densities in the magnetic circuit shown are 1·48 T, 0·808 T and 0·133 T respectively.

Solution by the ampere–turn formula. Let B_A T be the flux density in the limb A. Then

$$IT_A = \frac{B_A \times 400 \times 10^{-3}}{1.257 \times 10^{-6} \times 2000} = \frac{200 B_A}{1.257} \text{ A}$$

Since limb C is in parallel with limb A then the ampere–turns for A will also produce flux in C. Let B_C T be the flux density in the limb C. Then

$$\frac{200 B_A}{1.257} = \frac{B_C \times 400 \times 10^{-3}}{1.257 \times 10^{-6} \times 2000} + \frac{B_C \times 2 \times 10^{-3}}{1.257 \times 10^{-6}}$$

$$= \frac{2200}{1.257} B_C$$

$$B_C = \frac{200 \times 1.257}{1.257 \times 2200} B_A = \frac{B_A}{11} \text{ T.}$$

Total flux in A $= B_A \times 600 \times 10^{-6} \text{ Wb}$

Total flux in C $= \dfrac{B_A}{11} \times 600 \times 10^{-6} \text{ Wb}$

Total flux in B $= 600 \times 10^{-6} \left[B_A + \dfrac{B_A}{11} \right] \text{ Wb}$

$$= \frac{12 B_A}{11} \times 600 \times 10^{-6} \text{ Wb}$$

Flux density in $B = \dfrac{12 B_A \times 600 \times 10^{-6}}{11 \times 1.2 \times 10^{-3}} \text{ T} = \dfrac{6 B_A}{11} \text{ T}$

Hence

$$IT_B = \frac{6B_A \times 200 \times 10^{-3}}{11 \times 1\cdot257 \times 10^{-6} \times 2000} \, A$$

$$= \frac{600B_A}{11 \times 1\cdot257} \, A$$

Total m.m.f. $\quad = IT_A + IT_B$

$$300 = \frac{200B_A}{1\cdot257} + \frac{600B_A}{11 \times 1\cdot257}$$

$$= \frac{200B_A}{1\cdot257}\left[1 + \frac{3}{11}\right]$$

$$= \frac{200B_A \times 14}{1\cdot257 \times 11}$$

$$B_A = \frac{300 \times 1\cdot257 \times 11}{200 \times 14}\,T = 1\cdot48 \, T$$

Hence

Flux density in $B = \dfrac{6 \times 1\cdot48}{11}\,T = 0\cdot808 \, T$

Flux density in $C = \dfrac{1\cdot48}{11}\,T = 0\cdot134 \, T$

The flux densities produced are 1·48 T, 0·808 T and 0·134 T respectively.

Example 4.4

A magnetic circuit consists of two reluctances in series. If the value of one of these reluctances is $1\cdot8 \times 10^6$ SI units estimate the value of the second reluctance when an m.m.f. of 400 A applied to the circuit produces a flux of 0·2 mWb.

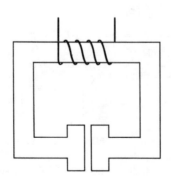

Fig. 4.3 Magnetic circuit for Example 4.5.

Example 4.5
Fig. 4.3 shows a magnetic circuit, the details of each part being as follows:

   Yoke: length 300 mm, area 400 mm², relative permeability 1000.
   Poles: length 50 mm/pole, area 800 mm², relative permeability 1500.
   Air gap: length 2 mm.

Determine the maximum value of m.m.f. which may be applied to the circuit so that the flux density in any one part of the circuit does not exceed 1 T.

Example 4.6
Neglecting the reluctance of all magnetic material in a magnetic circuit, the circuit may be considered as two reluctances in parallel. The flux across one air gap, which has a reluctance of 2×10^6 SI units, is 0·3 mWb. If the total flux in the circuit is 0·8 mWb determine the reluctance of the second gap.

Example 4.7
Neglecting all reluctances except those of two air gaps, which may be considered in parallel, an m.m.f. of 600 A produces a total flux of 1 mWb. One of the gaps has a reluctance of 10^6 SI units. Calculate the reluctance of the other gap and determine how the flux will divide between the two branches.

Example 4.8
A transformer core consists of a double magnetic circuit with the magnetising winding on the centre limb, which has a reluctance of 20×10^3 SI units and the reluctances of the outer limbs are of 40×10^3 and 50×10^3 SI units respectively. The maximum flux, in either of the outer limbs is limited to 8 mWb. Calculate the total m.m.f. required.

Example 4.9
A flux of 0·5 mWb has to be produced in the 2 mm air gap of a magnetic circuit having an iron path of 785 mm and a cross-sectional area of 700 mm². The iron, which has a relative permeability of 1020 at the working flux density, is wound with 600 turns of wire. Calculate the necessary value of magnetizing current.

Example 4.10
Fig. 4.4. is a dimensioned drawing of a symmetrical magnetic circuit. The air gap has a length of 1·5 mm and the iron has a relative permeability of 1500 at the working flux density. Determine the m.m.f. of the coil on the centre limb needed to give a total flux of 4 mWb in the gap.

Example 4.11
A horse-shoe electro-magnet has a magnetic length of 400 mm and a cross-sectional area of 1200 mm². If there is an effective air gap of 0·75 mm between

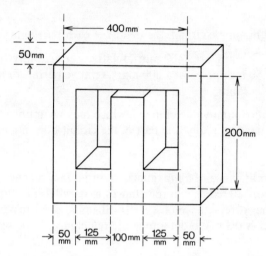

Fig. 4.4 Magnetic circuit, with dimensions, for Example 4.10.

each limb of the magnet and a steel keeper across the ends of the magnet, calculate the total flux produced by an m.m.f. of 800 A applied to the iron, which may be assumed to have a relative permeability of 500. Neglect the reluctance of the keeper.

Example 4.12
An iron rod, 300 mm long and having a cross-sectional area of 1000 mm², is bent into a toroid but with the two ends kept apart by a 2 mm brass spacer. When an m.m.f. of 690 A is applied to the iron, the total flux in the brass is 0·4 mWb. Estimate the relative permeability of the iron under these conditions.

Example 4.13
Fig. 4.5 shows one type of overload protection mechanism, which is set to operate at 20 A. An average flux density of 0·1 T is required in the air gap,

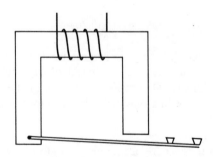

Fig. 4.5 Overload coil, reference Example 4.13.

which has an average length of 4 mm and the flux path through the iron has a length of 60 mm. Assuming a relative permeability of 500 for the iron, estimate, to the nearest turn, the number of turns required in the coil. Neglect the reluctance of the moving arm.

4.2 The Use of Magnetic Curves

Since the relative permeability of magnetic materials varies, not only with the material, but for any one material with the flux density in the material and the temperature of the material, it is an unjustifiable assumption to use one value for this permeability when different sections of the material may have different flux densities. Hence it is preferable to use part of the B/H curve for the material, or part of the μ_r/B curve.

Example 4.14
Points on the B/H curve for a magnetic material are given below:

B in T	1·04	1·15	1·2
H in A/m	1000	1450	1800

Estimate the m.m.f. required to produce a flux density of 1·1 T in a 250 mm length of the material.

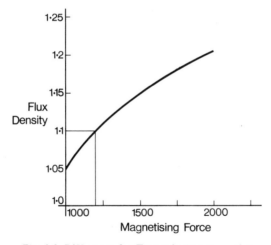

Fig. 4.6 B/H curve for Example 4.14 (to scale).

Fig. 4.6 shows the curve drawn through the given points. The scales used are as follows:

Vertical scale: flux density B 10 mm $\equiv 0\cdot05$ T

Horizontal scale: magnetizing force H 10 mm $\equiv 250$ A/m

63

Mark the curve with the required value of B, i.e. 1·1 T and so obtain the corresponding value of H.

From the curve, when $B = 1·1$ T, $H = 1200$ A/m. But m.m.f. $= Hl$ A and so

$$\text{mmf} = 1200 \times 250 \times 10^{-3} \text{ A} = 300 \text{ A}$$

The required m.m.f. is 300 A.

Note: Except in a very few cases it is not necessary to give the m.m.f. to a greater accuracy than the nearest ampere. In most cases the nearest 10 A will be sufficient.

Example 4.15

It is required to produce a flux of 0·5 mWb in the 2 mm air gap of a magnetic circuit which has an iron path of 800 mm and an area of 750 mm². Points on the B/H curve for the iron are given below:

B in T	0·6	0·7	0·8
H in A/m	478	518	580

If the iron is uniformly wound with 500 turns of wire, calculate the value of current required.

Since the total flux is $0·5 \times 10^{-3}$ Wb, therefore

$$\text{flux density } B = \frac{0·5 \times 10^{-3}}{750 \times 10^{-6}} \text{T} = 0·67 \text{ T}$$

Since
$$IT = Bl/\mu_0 \mu_r, \text{ for the air gap}$$

$$IT = \frac{0·67 \times 2 \times 10^{-3}}{1·257 \times 10^{-6} \times 1} \text{ A} = 960 \text{ A}$$

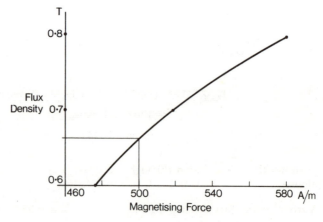

Fig. 4.7 *B/H* curve for Example 4.15 (to scale).

Fig. 4.7 shows the relevant part of the B/H curve for the iron, plotted to the following scales:

Vertical scale: flux density B \qquad\qquad 10 mm $\equiv$ 0·05 T

Horizontal scale: magnetizing force H \quad 10 mm $\equiv$ 20 A/m

From the curve, when $B = 0·67$ T, $H = 502$ A/m. Hence for iron

$$IT = Hl = 502 \times 800 \times 10^{-3} \text{ A}$$
$$= 401·6 \text{ A}$$
$$\text{Total m.m.f.} = 960 + 400 \text{ A} = 1360 \text{ A}$$

Hence
$$I = \frac{1360}{500} \text{ A} = 2·72 \text{ A}$$

The required current is 2·72 A.

Example 4.16
If in Example 4.14 an m.m.f. of 375 A had been applied to the iron, estimate the flux density which would have been produced in the magnetic material.

Example 4.17
A transformer core has a length of 0·4 m and a cross-sectional area of $6·4 \times 10^{-3}$ m^2. Determine the m.m.f. required to produce a flux of 5·8 mWb in the core. Points on the B/H curve for transformer steel are:

B in T	0·6	0·85	1·0	1·2
H in A/m	80	160	272	520

Example 4.18
A 200 mm length of Stalloy (silicon steel) has to undergo cycles of magnetization and the relative permeability of Stalloy varies with the flux density according to the following values:

B in T	0·3	0·5	0·8	0·9
μ_r	5750	6500	5100	4500

Calculate the magnetizing force when the flux density in the Stalloy is 0·65 T.

Example 4.19
In a two-pole d.c. machine the mean length of the flux path through the yoke and the poles is 600 mm and 5 mm through the air gaps. The path through the armature may be neglected. Calculate the m.m.f. required to produce a flux of 9 mWb in the air gap under the pole face of area 50×10^{-3} m^2. Points on the B/H curve are:

Q

B in T	0·1	0·15	0·20	0·25
H in A/m	800	1120	1760	2720

4.3 Magnetic Leakage and Fringing

Fig. 4.8 shows the practical effect of introducing an air gap into a magnetic circuit. Some of the flux in the magnetic material will leave the material

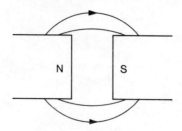

Fig. 4.8 Diagram of magnetic field showing leakage.

before the boundary of the air gap is reached. This is referred to as magnetic leakage.

Fig. 4.9 shows that the flux which does leave the magnetic material at the boundary of the air gap extends beyond the area of the gap subtended by the magnetic material. This is referred to as fringing.

In order then that a certain value of flux density shall be produced in the

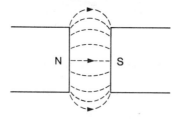

Fig. 4.9 Diagram of magnetic field showing fringing.

air gap, the area of which is always taken as being equal to the area of the poles, it is necessary to produce in the magnetic material a greater value of flux than is required in the air gap. The ratio between the total flux in the magnetic material to the total flux in the air gap is called the *leakage co-efficient*, and its value may vary between 1 and 1·2 depending on the circuit.

Example 4.20
An iron ring has a uniform cross-section of 1000 mm² and a mean circum-

ference of 1 m. A current of 1·82 A flowing in a 500 turn coil, wound uniformly on the ring, produces a flux of 1 mWb in the ring. If a 2 mm cut is made in the ring, what current is required to produce a flux of 1 mWb in the gap, assuming that the relative permeability of the iron remains constant, and allowing for a leakage coefficient of 1·2?

Without gap.

$$\text{Total flux } \Phi = \mu_0 \mu_r ITA/l$$

Hence

$$\mu_r = \Phi l/\mu_0 ITA$$

$$= \frac{1 \times 10^{-3} \times 1}{1\cdot257 \times 10^{-6} \times 1\cdot82 \times 500 \times 1000 \times 10^{-6}}$$

$$= 876$$

With gap.

$$\text{Flux density } B = \frac{1 \times 10^{-3}}{1000 \times 10^{-6}} = 1 \text{ T}$$

For air gap,

$$IT = \frac{1 \times 2 \times 10^{-3}}{1\cdot257 \times 10^{-6} \times 1} A = 1590 \text{ A}$$

For iron,

$$B = \text{flux density in air} \times \text{leakage coefficient}$$

$$= 1 \times 1\cdot2 \text{ T}$$

Hence

$$IT = \frac{1\cdot2 \times 1}{1\cdot257 \times 10^{-6} \times 876} = 1090 \text{ A}$$

$$\text{Total m.m.f.} = 1590 + 1090 \text{ A} = 2680 \text{ A}$$

$$\text{Current required} = \frac{2680}{500} \text{ A} = 5\cdot36 \text{ A}$$

The value of current required is 5·36 A.

Example 4.21
A total m.m.f. of 1200 A is applied to an electro-magnetic system, in which the length of the magnetic material is 500 mm, and in which there is an effective air gap of 1·5 mm. Allowing a leakage coefficient of 1·1, calculate the flux density produced in the air gap if the magnetic material, under the working conditions, has a relative permeability of 2500.

Let *B* be the flux density in the air gap. Then flux density in the magnetic material is 1·1 *B*. Summing the *AT* for each section of the system,

67

$$1200 = \frac{1 \cdot 1B \times 500 \times 10^{-3}}{1 \cdot 257 \times 10^{-6} \times 2500} + \frac{B \times 1 \cdot 5 \times 10^{-3}}{1 \cdot 257 \times 10^{-6} \times 1}$$

$$= \frac{B \times 10^{-3}}{1 \cdot 257 \times 10^{-6}} \left[\frac{1 \cdot 1 \times 500}{2500} + \frac{1 \cdot 5}{1} \right]$$

$$= \frac{10^3 B}{1 \cdot 257} (0 \cdot 22 + 1 \cdot 5) = \frac{10^3 B \times 1 \cdot 72}{1 \cdot 257}$$

Hence $\quad B = \dfrac{1200 \times 1 \cdot 257}{1720} \, \text{T} = 0 \cdot 876 \, \text{T}$

The flux density produced in the air gap is 0·876 T.

Example 4.22

A magnetic circuit, of cross-sectional area 10^3 mm^2, has an iron length of 1 m and a 2 mm air gap. The iron is wound with 1000 turns and when the coil carries a current of 3·05 A the flux in the air gap is 1·5 mWb. Points on the B/H curve for the iron are as follows:

B in T	1·4	1·6	1·8
H in A/m	576	635	765

Estimate the value of the leakage coefficient.

Fig. 4.10 shows the B/H curve

Total m.m.f. $\quad = 3 \cdot 05 \times 1000 \, \text{A} = 3050 \, \text{A}$

Flux density in air gap $= \dfrac{1 \cdot 5 \times 10^{-3}}{10^{-3}} \, \text{T} = 1 \cdot 5 \, \text{T}$

m.m.f. for air gap $\quad = \dfrac{1 \cdot 5 \times 2 \times 10^{-3}}{1 \cdot 257 \times 10^{-6}} \, \text{A} = 2390 \, \text{A}$

m.m.f. for iron $\quad = \text{total m.m.f.} - \text{m.m.f. for air gap}$

$\quad = 3050 - 2390 \, \text{A} = 660 \, \text{A}$

The magnetizing force for iron is 660×1 A/m.

From the curve shown in Fig. 4.10, corresponding to a magnetizing force of 660 A/m, the flux density is 1·65 T.

$$\text{Leakage coefficient} = \frac{\text{flux density in the iron}}{\text{flux density in the air gap}}$$

$$= \frac{1 \cdot 65}{1 \cdot 5} = 1 \cdot 1$$

The leakage coefficient for the magnetic circuit is 1·1.

Example 4.23

An electro-magnet, having a total length of 2 m and a cross-sectional area of

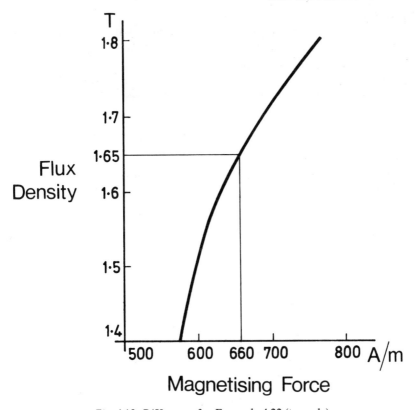

T

1·8

1·7

1·65

Flux
Density   1·6

1·5

1·4

500   600   660 700   800 A/m

Magnetising Force

Fig. 4.10 *B/H* curve for Example 4.22 (to scale).

2000 mm², is wound with 1000 turns. Between the magnet and the load
there is an effective air gap of 2 mm. Calculate the magnetizing current
required to produce a flux of 2·4 mWb in the air gap, assuming a leakage
coefficient of 1·2. Points on the *B/H* curve are as follows:

B in T	1·2	1·3	1·4	1·5
H in A/m	500	563	655	775

Example 4.24
An iron ring of mean diameter 200 mm and cross-sectional area of 10^{-3} m²
is uniformly wound with 500 turns of wire. Allowing a leakage coefficient
of 1·2, determine the magnetizing current required to produce a flux of
1 mWb in a 2 mm air gap cut in the iron. Points on the *B/H* curve are:

B in T	0·9	1·1	1·3	1·4
H in A/m	470	500	550	600

69

Example 4.25

An iron ring of mean circumference 0·75 m and of 500 mm^2 cross-sectional area is uniformly wound with 500 turns of wire. When a current of 1·5 A is maintained in the coil, a flux of 0·6 mWb is produced in the iron. Calculate the relative permeability of the iron. Assuming that the relative permeability of the iron remains constant, determine the flux density in a 2 mm gap cut into the ring if the magnetizing current is increased to 4 A. Allow a leakage coefficient of 1·2.

Example 4.26

Calculate the flux in a 1·5 mm air gap cut into an iron ring, of mean diameter 125 mm and having a cross-sectional area of 300 mm^2, when a total m.m.f. of 2000 A is applied to the magnetic circuit. Allow a leakage coefficient of 1·1 and assume a relative permeability of 1000 for the iron at the working flux density.

Example 4.27

A magnetic circuit consists of an iron path, 0·6 m in length and 500 mm^2 in cross-sectional area, and an effective air gap of length 2 mm. If a total m.m.f. of 2·8 × 10^3 A is required to produce a flux of 0·6 mWb in the air gap, estimate the relative permeability of the iron, at the working flux density, if there is a leakage coefficient of 1·2.

Example 4.28

Determine the leakage coefficient for a magnetic circuit, which requires a total m.m.f. of 950 A to produce a flux density of 0·5 T in a 1 mm air gap, when the magnetic path consists of 200 mm of cast iron. Part of the B/H curve for the iron is given by the following points:

B in T	0·3	0·48	0·6	0·75
H in A/m	1000	2000	3300	6000

Example 4.29

A flux density of 0·7 T is produced in the 2·5 mm air gap in a magnetic circuit when an m.m.f. of 1500 A is applied to the circuit. Given points on the B/μ_r curve for the magnetic material in the circuit, and allowing a leakage coefficient of 1·2, estimate the effective length of magnetic material in the circuit.

B in T	0·5	0·7	0·9	1·0
μ_r	4250	4750	4700	4250

Chapter 5

Electromagnetic Induction

5.1 Faraday's and Lenz's Laws

Faraday's Law states that when a conductor cuts or is cut by a magnetic flux, an e.m.f. is induced in the conductor and the magnitude of the e.m.f. is proportional to the rate of cutting flux. Lenz's Law is concerned with the direction of the induced e.m.f. It states that the direction of this e.m.f. around a closed circuit is always such as to produce a force which opposes the force producing the relative motion between the conductor and the flux. The induced e.m.f. always acts in that direction so as to reduce the rate of cutting flux.

Example 5.1
A conductor of length 0·5 m is moved at a constant velocity of 4 m/s in a direction perpendicular to a uniform magnetic field of density 1·1 T. Calculate the e.m.f. induced in the conductor when it has moved a distance of 1 m.

$$\text{Area of flux cut} = \text{length of conductor} \times \text{distance moved}$$
$$= 0·5 \times 1 \text{ m}^2 = 0·5 \text{ m}^2$$

and so   total flux cut $= 1·1 \times 0·5 \text{ Wb} = 0·55 \text{ Wb}$

But the time taken to move a distance of 1 m is 0·25 s ($V = 4$ m/s). Hence

$$\text{flux cut per second} = \frac{0·55}{0·25} \text{ Wb/s} = 2·2 \text{ Wb/s}$$

But   flux cut per second $=$ induced e.m.f.

and so   e.m.f. induced in conductor $= 2·2$ V

Example 5.2
A conductor of length 0·4 m is moved at a constant velocity of 1 m/s at an angle to a uniform field of density 1·2 T. If the e.m.f. induced in the conductor is 240 mV, determine the angle which the direction of motion of the conductor makes with the field.

Let $\Theta°$ be the angle the conductor's path makes with the field. Then the component of velocity perpendicular to the flux is

$$V \sin \Theta = 1 \times \sin \Theta$$

Hence the induced e.m.f. is proportional to $\sin \Theta$, and therefore

$$\begin{aligned}
\text{area of flux cut/s} &= \sin \Theta \times 0.4 \, \text{m}^2/\text{s} \\
\text{flux cut/s} &= 1.2 \times 0.4 \sin \Theta \, \text{Wb/s} \\
\text{induced e.m.f.} &= 0.48 \sin \Theta \, \text{V}
\end{aligned}$$

But

$$\text{induced e.m.f.} = 240 \times 10^{-3}$$

and so

$$0.48 \sin \Theta = 240 \times 10^{-3}$$

$$\sin \Theta = \frac{0.24}{0.48} = 0.5$$

$$\Theta = 30°$$

The conductor must be moving at an angle of $\pi/6$ rad to the field.

Example 5.3
A search coil of 20 turns is wound over the turns of an iron-cored solenoid. If the flux in the solenoid changes from 2·0 mWb to 4·5 mWb in 0·2 s, determine the average e.m.f. induced in the search coil.

$$\text{Change in flux} = 4.5 - 2.0 \, \text{mWb} = 2.5 \, \text{mWb}$$

The time taken for this change is 0·2 s, and so

$$\text{average rate of change of flux} = \frac{2.5 \times 10^{-3}}{0.2} \, \text{Wb/s}$$

$$= 12.5 \times 10^{-3} \, \text{Wb/s}$$

$$\text{average e.m.f./turn} = 12.5 \, \text{mV}$$

$$\text{average e.m.f. induced in coil} = 20 \times 12.5 \, \text{mV} = 250 \, \text{mV}$$

The average e.m.f. induced in the search coil is 250 mV or 0·25 V.

Example 5.4
A coil of 10 turns, on the rotor of a 4-pole d.c. machine, has an angular velocity of 250 rad/s. If the flux/pole is 10 mWb, determine the average e.m.f. induced in the coil.

In 2π rad, 1 side of 1 turn will cut $4 \times 10 \times 10^{-3}$ Wb, and so in 2π rad, 1 turn will cut $2 \times 40 \times 10^{-3}$ Wb. But the coil moves through 250 rad/s. Hence

$$\text{flux cut/turn/s} = \frac{80 \times 10^{-3} \times 250}{2\pi} \, \text{Wb/s} = 3.18 \, \text{Wb/s}$$

e.m.f. induced/turn $\quad = 3·18$ V

e.m.f. induced in coil $= 3·18 \times 10$ V $= 31·8$ V

The average e.m.f. induced in the coil will be 31·8 V.

Example 5.5

The rotor, of a 4-pole d.c. machine, carries a total of 500 conductors, in two parallel paths. Determine the angular velocity of the rotor so that, when the flux/pole is 8 mWb the generated e.m.f. is 240 V.

e.m.f. of machine $\quad\quad = $ e.m.f. induced/path

No. of conductors/path $= \dfrac{500}{2} = 250$

In 2π rad, 1 conductor cuts $4 \times 8 \times 10^{-3}$ Wb, and so in 2π rad, 250 conductors will cut $250 \times 32 \times 10^{-3}$ Wb $= 8$ Wb.

Let ω rad/s be the angular velocity of the rotor.

induced e.m.f./path $= 8\omega/2\pi$ V

But $\quad\quad$ induced e.m.f./path $= 240$ V

Hence $\quad\quad 4\omega/\pi = 240$ V

$\omega = 60\pi$ rad/s

$= 188$ rad/s

The angular velocity of the rotor has to be 188 rad/s.

Example 5.6

A conductor, 600 mm long, is moved with a constant velocity of 2·5 m/s at right angles to a uniform flux of density 0·8 T. Calculate the e.m.f. induced in the conductor.

Example 5.7

When a conductor of length 0·4 m is moved with a constant velocity of 1·5 m/s at an angle of 45° to a uniform magnetic field, the e.m.f. induced in the conductor is 250 mV. Estimate the flux density of the field.

Example 5.8

A conductor of length 0·75 m is moved with a constant velocity at an angle of 60° to a uniform field of density 1·2 T. If the e.m.f. induced in the conductor is 0·68 V determine the velocity of the conductor.

Example 5.9

An iron-cored solenoid, wound with 200 turns, has a length of 250 mm and a cross-sectional area of 500 mm². If a search coil of 10 turns is wound over the solenoid, calculate the e.m.f. induced in the search coil when the current in the solenoid increases from 0·1 A to 0·5 A in 2 s. Points on the B/H curve for the core of the solenoid are as follows:

B in T	0·6	0·85	1·0	1·2	1·4
H in A/m	80	160	272	520	1120

Example 5.10

A conductor of length 0·4 m carries a current of 5 A and lies at right angles to a magnetic field of density 0·9 T. Calculate the force acting on the conductor and the power required to move the conductor at a constant velocity of 1·5 m/s in a direction perpendicular to the field. Calculate the e.m.f. induced in the conductor and hence show that the power acting on the conductor is equal to the electrical power produced.

Example 5.11

The rotor of a 4-pole d.c. machine, carries 480 conductors connected in four parallel paths, and has an angular velocity of 50π rad/s. Estimate the e.m.f. generated by the machine when the flux/pole is 20 mWb.

Example 5.12

When a 2-pole d.c. generator is supplying a current of 12 A the mechanical power input to the generator is 3·4 kW. If the efficiency of the generator is 84 per cent, estimate the angular velocity at which the machine is being driven. The rotor of the generator carries 200 conductors/path and the flux/pole is 25 mWb. Neglect any voltage drops in the generator.

5.2 Self-induced e.m.f.

Whenever there is a change in the magnitude of a current in a conductor or in a circuit there is an accompanying change in the flux associated with the current. Hence there will be relative motion between the flux and the conductor and so an e.m.f. will be induced in the conductor.

This e.m.f. is known as the e.m.f. of self induction (e_L) and its magnitude depends on the rate of change of the current and that property of the conductor or of the circuit, known as the inductance (L) of the conductor or circuit.

Inductance is that property of a circuit which opposes any change in the value of the current in that circuit and the unit of inductance is the Henry (H). A circuit is said to have an inductance of 1 H when the current changing at the rate of 1 A/s induces in the circuit an e.m.f. of 1 V. Hence

$$e_L = L \times \text{rate of change of current}$$

$$= L\frac{i_2 - i_1}{t_2 - t_1}$$

As the e.m.f. of self inductance is an induced e.m.f. it must obey both Faraday's and Lenz's Laws.

$$e_L = T \times \frac{\phi_2 - \phi_1}{t_2 - t_1}$$

Example 5.13
In an inductive circuit, wound with 400 turns, the current increases from zero to 5 A in 0·4 s and the accompanying change in flux is from zero to 0·50 mWb in the same time. Calculate the inductance of the circuit.

Since
$$e_L = L \times \frac{i_2 - i_1}{t_2 - t_1}$$

and
$$e_L = T \times \frac{\phi_2 - \phi_1}{t_2 - t_1}$$

then
$$L \times \frac{5 - 0}{0·4 - 0} = 400 \times \frac{(0·5 - 0) 10^{-3}}{0·4 - 0}$$

$$L = \frac{400 \times 0·5 \times 10^{-3}}{5} \text{ H}$$

$$= 0·04 \text{ H}$$

The inductance of the circuit is 0·04 H or 40 mH.
    Note: From Example 5.13 it follows that

$$L = T\Phi/I$$

The product of the flux and the number of turns is called the flux linkages.

$$L = \text{flux linkages}/A$$

Example 5.14
A coil of 150 turns is wound on a bar of iron 250 mm in length and 40 mm in diameter. If the relative permeability of the iron, at the working flux density, is 2000, estimate the inductance of the coil.
    Neglecting the diameter of the wire, the area of the field is given by

$$A = \frac{\pi \times 40^2}{4} \text{ mm}^2$$

Let the current in the coil be I. Then

$$\Phi = \mu_0 \mu_r IT A/l$$

$$\frac{\Phi}{I} = \frac{1·257 \times 10^{-6} \times 2000 \times 150\pi \times 40^2 \times 10^{-6}}{4 \times 250 \times 10^{-3}} \frac{\text{Wb}}{\text{A}}$$

$$L = \frac{\Phi T}{I} = \frac{1·257 \times 2 \times 15 \times \pi \times 16 \times 150 \times 10^{-6}}{1} \text{H}$$

$$= 284 \text{ mH}$$

The inductance of the coil is 284 mH or 0·28 H.

Note: From Example 5.14

$$L = \frac{\mu_0 \mu_r T^2 A}{l}$$

$$= \frac{T^2}{l/\mu_0 \mu_r A} = \frac{T^2}{S},$$

where S is the reluctance of the magnetic circuit.

Example 5.15

Given that $L = \mu_0 \mu_r T^2 A/l$, determine the fundamental unit for the permeability of free space.

Since

$$L = \frac{\mu_0 \mu_r T^2 A}{l}$$

then

$$\mu_0 = \frac{Ll}{\mu_r T^2 A} \text{ SI units}$$

Substituting the fundamental units for these quantities,

$$\mu_0 = \frac{H \times m}{m^2} = \frac{H}{m}$$

(μ_r and T have no units).

Hence the fundamental unit for the permeability of free space is the henry/metre. Its magnitude is $1·257 \times 10^{-6}$.

$$\mu_0 = 1·257 \times 10^{-6} \text{ H/m}$$

Example 5.16

A coil has an inductance of 0·1 H and carries a current of 20 A. Calculate the average e.m.f. induced in the coil if the current is completely reversed in 3·0 s.

$$\text{Change in current} = i_2 - i_1$$

$$= -20 - 20 = -40 \text{ A}$$

(Since only the magnitude of the induced e.m.f. is asked for in the question, the negative sign for the current change may be disregarded.)

Average rate of change of current $= 40/3$ A/s

But

average induced e.m.f. $= L \times$ rate of change of current

$$= 0·1 \times \frac{40}{3} \text{ V} = 1·33 \text{ V}$$

The average e.m.f. of self inductance is 1·33 V.

Had the direction of the self-induced e.m.f. been required, it would have acted in that direction to oppose the reversal of current, i.e. it would act in the same direction as that in which the original current was flowing.

Example 5.17
An inductive coil carries a current of 5 A. When the current is reduced to zero in 0·25 s the induced e.m.f. is 25 V. Estimate the inductance of the coil.

Example 5.18
When a current of 20 A, flowing in a circuit having an inductance of 0·2 H, is uniformly reduced to zero the resulting e.m.f. is 20 V. Calculate the time taken for the reduction of the current.

Example 5.19
A coil of 0·3 H inductance carries a unidirectional current. When this current is reduced to zero in 1·5 s the self induced e.m.f. is 1·2 V. Determine the value of the original current.

Example 5.20
A magnetizing current of 5 A, flowing in a coil of 200 turns, produces a flux of 5 mWb. What is the inductance of the coil?

Example 5.21
A coil with a non-magnetic core is wound in the form of a closed ring with a mean diameter of 200 mm and a circular cross section of 50 mm diameter. If the coil consists of 500 turns of wire, 1 mm in diameter, calculate the inductance of the coil.

Example 5.22
In an air-cored coil of 200 turns a flux of 1 μWb is associated with a current of 10 A and when this current is uniformly reduced to zero the e.m.f. induced in the coil is 2 mV. Calculate the time taken for the current to decay to zero.

Example 5.23
A coil of 800 turns is wound on a hollow cylindrical bakelite former of 70·7 mm external diameter and a current of 5 A through the coil produces an average flux density of 5 mT. Calculate the inductance of the coil and the average e.m.f. induced in the coil if the current is completely reversed in 2 s.

Example 5.24
Given that the permeability of free space is $1·257 \times 10^{-6}$ H/m determine the fundamental unit for magnetic reluctance.

77

5.3 Current and Voltage/Time Curves

When a d.c. voltage is applied to an inductive circuit the current does not immediately reach its final value but takes a finite time to reach this value: this time depends on the relative values of inductance and resistance.

The rate of growth of the current is not a constant but decreases as the value of the current in the circuit increases. If the original rate of change of current were to remain constant then the final value of current would be reached in time T s. This interval of time is called the *time constant* $(T = L/R)$.

Example 5.25
A 240 V d.c. supply is switched on to a coil having a resistance of 120 Ω and

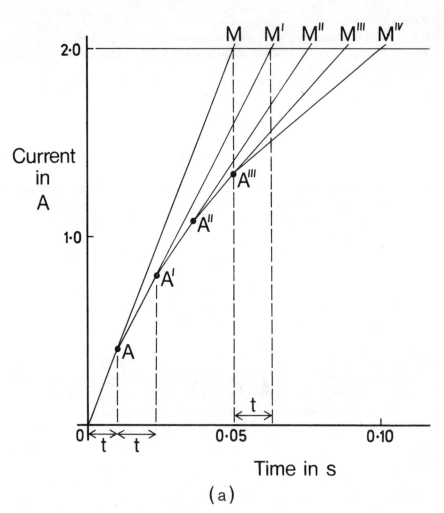

(a)

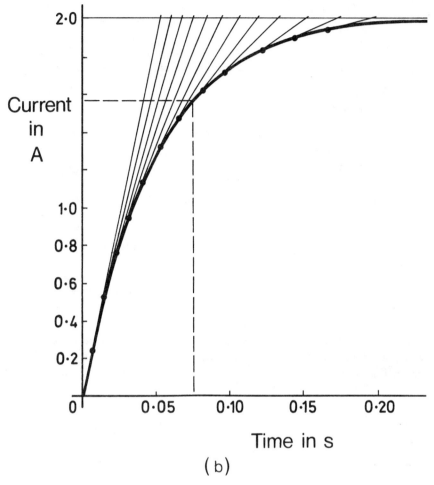

Fig. 5.1 (a) Construction of a current/time curve for $L - R$ circuit (to scale).
(b) Curve for Example 5.25 (to scale).

an inductance of 6 H. Draw the approximate current/time curve and hence obtain the value of the volt drop across the resistance after a time period of 75 ms. Determine the Ohm's law value of current and the time constant for the circuit.

$$\text{Final value of current} = \frac{V}{R} = \frac{240}{120}\text{A}$$

$$= 2\text{ A}$$

79

$$\text{Time constant} \quad = \frac{L}{R} = \frac{6}{120}\text{ s}$$

$$= 50\text{ ms}$$

Fig. 5.1(a) shows the method used to draw the curve. Choose suitable scales:

Vertical scale: Instantaneous value of current 1 mm ≡ 0·02 A

Horizontal scale: Time   1 mm ≡ 1·25 ms

Draw a horizontal line through $I = 2$ A and along this line mark a point, M, distant 50 ms (the time constant) from the vertical axis.

Draw a straight line through the origin O and the point M. This line represents the growth of current if the initial rate of change of current were to remain constant. Assume that this rate of change of current is maintained till the point A is reached. The rate of change then alters.

Measure the time taken (t s) for the current to reach point A and mark a point M' along the 2 A horizontal such that MM' equals t s. Join A and M' with a straight line.

The horizontal projection of this line, along the horizontal through A, will be equal to the time constant (50 ms). Along AM' mark another point A' and repeat the above construction. Repeat for a number of such points over a period of time equal to 4 × time constant (200 ms).

Join all points like A, A', ... by a smooth curve. (The more points like A which are taken, i.e. the smaller the time interval t s, the more accurate will be the final curve.)

Fig. 5.1(b) shows the complete curve drawn to the same current scale as in Fig. 5.1(a) but with half the time scale, i.e. 1 mm ≡ 2·5 ms. From the curve, when the current has been increasing for 75 ms it has reached the value of 1·56 A. Hence

$$\text{volt drop across resistance} = iR = 1·56 \times 120\text{ V}$$

$$= 187\text{ V}$$

After 50 ms the value of current is 1·26 A. As a percentage of the final current that is

$$\frac{1·26}{2} \times 100\% = 63\%$$

After 75 ms the volt drop across the resistance will be 187 V and after 50 ms the current will be 63 per cent of the final current.

Note: After a time interval equal to the time constant of the circuit, the current and/or the voltage will have completed 63·2′ per cent of their total growth or decay. This provides a second means of defining the time constant of a circuit.

N.B. The curve for the decay of the current is drawn in the same manner but the starting point is the Ohm's Law value of current at a time of 0 s and the first straight line is drawn from this point to the time axis when $t = T$ (the time constant).

Example 5.26
A circuit has a resistance of 5 Ω and an inductance of 1·0 H. The current in this circuit is increased uniformly from 0 to 5 A in 0·1 s and is then increased uniformly from 5 A to 10 A in 0·2 s. The current is maintained constant at this value for a further 0·2 s and is then decreased uniformly to zero in 0·5 s. Draw graphs representing the variations of the current, the induced e.m.f. and the applied voltage to a time scale.

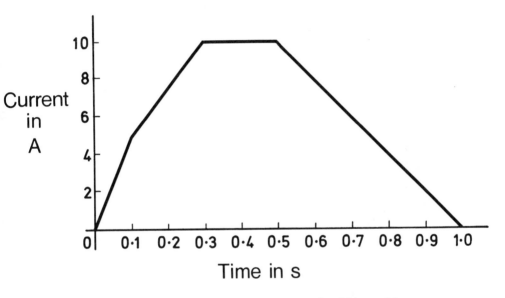

Fig. 5.2 Current/time curve for Example 5.26 (to scale).

Fig. 5.2 shows the current/time curve drawn to the following scales:

        Vertical scale: Current   1 mm ≡ 0·2 A
        Horizontal scale: Time   1 mm ≡ 0·01 s

Fig. 5.3 shows the curves for the voltage drops in the circuit, i.e. iR and e_L, drawn to the scales:

        Vertical scale: Voltage   1 mm ≡ 1 V
        Horizontal scale: Time   1 mm ≡ 0·01 s

To determine the curve for the resistive drop (iR), since the voltage drop

81

R

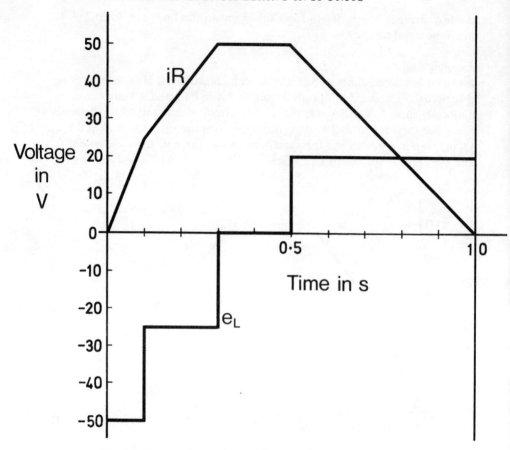

Fig. 5.3 Curves of e_L and iR against time for Example 5.26 (to scale).

across the resistance is proportional to the current then it will vary according to the following table:

Time in s	0–0·1	0·1–0·3	0·3–0·5	0·5–1·0
iR in V	0–25	25–50	50	50–0

To determine the curve for the induced e.m.f., $(L \times (i_2 - i_1)/(t_2 - t_1))$, since the current is varying uniformly then this curve will consist of a series of steps.

$$e'_L = 1 \times \frac{5 - 0}{0 \cdot 1 - 0} V = 50 \ V$$

As the current is increasing positively this will act in opposition, i.e. it will be negative.

$$e''_L = 1 \times \frac{10 - 5}{0 \cdot 3 - 0 \cdot 1} V = 25 \ V$$

This will also be negative.

When the current remains constant at 10 A there will be no induced e.m.f.

$$e_L''' = 1 \times \frac{0 - 10}{1 \cdot 0 - 0 \cdot 5} \, 5 \, \text{V} = 20 \, \text{V}$$

In this case the current is being decreased and therefore this e.m.f. of self inductance will act in the same direction, as did the original applied voltage.

Fig. 5.4 shows the applied voltage/time curve.

> Vertical scale: Voltage   1 mm ≡ 1 V
>
> Horizontal scale: Time   1 mm ≡ 0·01 s

Had there been no resistance in the circuit, the applied voltage curve would have been equal and opposite to the curve for the e.m.f. of self inductance. With resistance in the circuit then the curve of the applied voltage must be

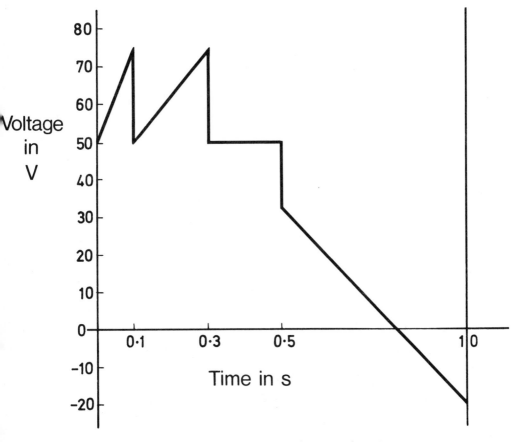

Fig. 5.4 Applied voltage/time curve for Example 5.26.

83

the sum of the curves for the iR drop and that which is equal and opposite to the curve for e_L.

Example 5.27
What is the time constant for a circuit consisting of a 98 Ω resistor in series with a coil of resistance 12 Ω and inductance 1·1 H?

Example 5.28
A coil has a resistance of 10 Ω and an inductance of 1·5 H. Calculate the series resistance required to reduce the time constant to 30 ms.

Example 5.29
Construct the current/time curve for the growth of current when a 240 V d.c. supply is switched on to a circuit having a resistance of 30 Ω and an inductance of 2·4 H. From this curve obtain the value of current (1) 50 ms and (2) 120 ms after making the circuit.

Example 5.30
A circuit, having a resistance of 40 Ω and an inductance of 1 H, carries a constant unidirectional current of 6 A. If, at the same time that the supply is removed, a short circuit is applied to the circuit, construct the voltage/time curve for the voltage across the resistance. From this curve obtain the voltage across the resistance 25 ms after the short circuit was applied. What is the value of current at this instant?

Example 5.31
The voltage, applied to a circuit of negligible resistance and 1·5 H inductance, is varied according to the following table:

Time interval in s	0–0·2	0·2–0·5	0·5–0·7	0·7–1·0
Voltage value in V	45	70	0	− 100

Draw the curves for the voltage and current/time. What is the maximum value of current and what is the current 0·4 s after the start of the voltage cycle? Assume that when the time is zero the current is zero.

Example 5.32
A circuit has a resistance of 10 Ω and an inductance of 2·0 H. The current in this circuit is increased uniformly from 0 A to 2 A in 0·2 s and is then increased uniformly from 2 A to 4 A in 0·3 s. The current remains constant at 4 A for a further 0·2 s and is then reduced uniformly to zero in 0·3 s. Draw graphs representing the variations of induced e.m.f. and the applied voltage to a time scale. From these curves obtain the maximum value of the applied voltage and the values of the induced e.m.f. and applied voltage when the current is 3 A and increasing.

5.4 Mutual Inductance

When two coils are so arranged, that the flux produced by one coil, in whole or in part, cuts the turns of the second coil, the two coils are said to possess *mutual inductance* (M). The unit of mutual inductance is the Henry.

Two coils are said to possess a mutual inductance of 1 H when the current in one coil, changing at the rate of 1 A/s induces in the second coil an e.m.f. of 1 V.

Example 5.33
Two coils A and B possess a mutual inductance of 0·45 H. Calculate the e.m.f. induced in coil B when the current in coil A is changing at the rate of 10 A/s.

From the definition of the unit of mutual inductance, for a rate of change of current of 1 A/s in coil A, the e.m.f. induced in coil B is given by

$$e_B = 0{\cdot}45 \ V$$

For a rate of change of 10 A/s,

$$e_B = 10 \times 0{\cdot}45 \ V = 4{\cdot}5 \ V$$

The e.m.f. induced in coil B will be 4·5 V.

Note: The e.m.f. of mutual inductance is

$$e_m = M \times \text{rate of change of current}$$

where M is the mutual inductance between the coils.

Example 5.34
Two coils A and B, of 300 and 200 turns respectively, are magnetically coupled. When the current in coil A changes from 1·5 to 2·0 A, the corresponding flux linked with the turns of coil B changes from 5·0 mWb to 6·0 mWb. Determine the mutual inductance between the two coils.

Since the e.m.f. produced in coil B is an induced e.m.f., it must obey Faraday's laws. Hence

$$e_B = T_2 \times \frac{\phi_2 - \phi_1}{t_2 - t_1}$$

where ϕ_2 and ϕ_1 are the values of flux at times t_2 and t_1 respectively. But

$$e_B = M \times \frac{i_2 - i_1}{t_2 - t_1}$$

and so

$$M \times \frac{i_2 - i_1}{t_2 - t_1} = T_2 \times \frac{\phi_2 - \phi_1}{t_2 - t_1}$$

$$M = T_2 \times \frac{\phi_2 - \phi_1}{i_2 - i_1}$$

$$= 200 \times \frac{(6 \cdot 0 - 5 \cdot 0)}{2 \cdot 0 - 1 \cdot 5} 10^{-3} \, \text{H}$$

$$= \frac{200 \times 1 \cdot 0 \times 10^{-3}}{0 \cdot 5} \, \text{H} = 0 \cdot 4 \, \text{H}$$

The mutual inductance between the two coils is 0·4 H.

Note: Let ϕ be the flux produced by the m.m.f. in coil A. Then

$$\phi = \frac{i_A T_1}{S}$$

where S is the reluctance of the common magnetic circuit.

Hence
$$\phi_2 - \phi_1 = \frac{i_2 T_1}{S} - \frac{i_1 T_1}{S} = \frac{T_1}{S}(i_2 - i_1)$$

then
$$\frac{\phi_2 - \phi_1}{i_2 - i_1} = \frac{T_1}{S}$$

$$M = \frac{T_1 T_2}{S}$$

Example 5.35

Two coils of 400 and 100 turns respectively are wound on a common magnetic circuit, which has a length of 0·8 m and a cross sectional area of 0·01 m². If the magnetic material has a relative permeability of 2500, at the working flux density, estimate the mutual inductance between the two coils.

Reluctance of the magnetic circuit $S = \dfrac{l}{\mu_0 \mu_r A}$

$$= \frac{0 \cdot 8}{1 \cdot 257 \times 10^{-6} \times 2500 \times 0 \cdot 01} \, \text{H}^{-1}$$

$$= 25 \cdot 4 \times 10^3 \, \text{H}^{-1}$$

But

mutual inductance M
$$= \frac{T_1 \times T_2}{S} = \frac{400 \times 100}{25 \cdot 4 \times 10^3} \text{H}$$

$$= 1 \cdot 57 \, \text{H}$$

The mutual inductance between the two coils will be 1·57 H.

Note: Since

$$M = \frac{T_1 \times T_2}{S}$$

then
$$M^2 = \frac{T_1^2 \times T_2^2}{S^2} = \frac{T_1^2}{S} \times \frac{T_2^2}{S}$$

$$= L_1 \times L_2$$
$$M = \sqrt{(L_1 \times L_2)}$$

Alternative solution using the relationship between the self inductances and the mutual inductance.

$$\text{Self inductance of first coil} \quad = \frac{400^2}{25 \cdot 4 \times 10^3} \text{H} = 6 \cdot 3 \text{ H}$$

$$\text{Self inductance of second coil} = \frac{100^2}{25 \cdot 4 \times 10^3} \text{H} = 0 \cdot 394 \text{ H}$$

then
$$M = \sqrt{(6 \cdot 3 \times 0 \cdot 394)} \text{ H} = 1 \cdot 57 \text{ H}$$

In practice the value of mutual inductance given by $\sqrt{(L_1 \times L_2)}$ will always be greater than the true value of M. This is because not all of the flux produced by the one coil will link with all the turns of the second coil.

The ratio $M/\sqrt{(L_1 \times L_2)}$ is the coupling factor, k. This coupling factor will always be less than one. When k approaches unity the coils are said to be tightly coupled, and they are said to be loosely coupled when k approaches zero.

Example 5.36
Two coils A and B are wound on a common magnetic circuit. When the current in A changes at the rate of 50 A/s, the e.m.f. induced in A is 10 V and that in B is 10·6 V. When the current in B changes at the rate of 25 A/s, the e.m.f. induced in B is 10 V. What is the coupling factor between the two coils?

Since the e.m.f. of self inductance $= L \times$ rate of change of current, then

$$e_A = 10 = L_A \times 50$$
$$L_A = \frac{10}{50} \text{H} = 0 \cdot 2 \text{ H}$$

Also
$$e_B = 10 = L_B \times 25$$
$$L_B = \frac{10}{25} \text{H} = 0 \cdot 4 \text{ H}$$

But e.m.f. of mutual inductance $= M \times$ rate of change of current in other coil, and so

$$e_M = 10 \cdot 6 = M \times 50$$
$$M = \frac{10 \cdot 6}{50} \text{H} = 0 \cdot 212 \text{ H}$$

But
$$M = k\sqrt{(L_A \times L_B)}$$

and so
$$k = \frac{0 \cdot 212}{\sqrt{(0 \cdot 2 \times 0 \cdot 4)}} = 0 \cdot 75$$

87

The coupling factor between the two coils is 0·75.

Note: From Example 5.36,

$$e_A \propto L_A \propto \frac{T_A^2}{S}$$

$$e_m \propto M \propto \frac{T_A T_B}{S}$$

$$\frac{e_M}{e_A} = \frac{T_A T_B}{S} \times \frac{S}{T_A^2} = \frac{T_B}{T_A}$$

This is the principle of the *transformer*, which is used in alternating current circuits for altering the value of voltages.

Example 5.37
A primary coil of 300 turns and a secondary coil of 200 turns are wound on a closed magnetic core. When a 240 V a.c. supply is applied to the primary the measured p.d. across the secondary is 144 V. If the primary has a self inductance of 0·9 H determine the coupling factor between the two coils and their mutual inductance.

With perfect coupling

$$e_S/e_P = T_S/T_P$$

$$e_S = 240 \times \frac{200}{300} \text{V} = 160 \text{ V}$$

But measured voltage is 144 V, and therefore

$$\text{coupling factor} = \frac{144}{160} = 0·9$$

With constant reluctance, $L \propto T^2$, but

$$\text{inductance of primary } (\propto 300^2) \quad = 0·9 \text{ H}$$

$$\text{inductance of secondary } (\propto 200^2) = 0·9 \times \frac{200^2}{300^2} \text{ H}$$

$$= 0·4 \text{ H}$$

$$M = k\sqrt{(L_P \times L_S)} \text{ H}$$

$$= 0·9\sqrt{(0·9 \times 0·4)} \text{ H}$$

$$= 0·9 \times 0·6 \text{ H} = 0·54 \text{ H}$$

With a coupling factor of 0·9 the mutual inductance between the windings is 0·54 H.

Alternative solution. Neglecting the resistance of the primary,

$$e_L = 240 \text{ V}$$

Hence $\qquad\qquad 240 = 0.9 \times$ rate of change of current

$\qquad$ Rate of change of current $= 240/0.9$ A/s

But

$$e_M = M \times \text{rate of change of current}$$

and so

$$M = 144 \times \frac{0.9}{240} \text{H} = 0.54 \text{H}$$

Coupling factor $k \qquad = \dfrac{0.54}{\sqrt{(L_P \times L_S)}}$

Since the windings are on the same magnetic circuit, then

$$L_S/L_P = 200^2/300^2$$

$$L_S = 0.9 \times \frac{200^2}{300^2} \text{H} = 0.4 \text{H}$$

$$k = \frac{0.54}{\sqrt{(0.9 \times 0.44)}} = \frac{0.54}{0.6}$$

$$= 0.9$$

For a mutual inductance of 0·54 H the coupling factor between the coils is 0·9.

Example 5.38
Two coils are magnetically coupled and when the current in one coil is changing at the rate of 20 A/s the e.m.f. induced in the second coil is 8 V. Calculate the mutual inductance between the coils.

Example 5.39
Two coils, wound on a common magnetic circuit, have a mutual inductance of 0·8 H. When the current in one of the coils is changing, the e.m.f. induced in the other coil is 20 V. At what rate must the current be changing?

Example 5.40
When the current in a coil, of 250 turns, changes from 0·8 A to 1·0 A the flux produced increases from 4 mWb to 6 mWb. Determine the mutual inductance between this coil and a second coil of 100 turns, so placed that all of the flux produced by the first coil cuts all of the turns of the second coil.

Example 5.41
Two coils of 300 and 100 turns respectively, wound on a common magnetic circuit have a mutual inductance of 1 H. Assuming perfect coupling, calculate the reluctance of the magnetic circuit.

Example 5.42
A coil of 400 turns has an inductance of 0·8 H. Assuming perfect coupling between this coil and a second coil of 250 turns, determine the mutual inductance between the two coils.

Example 5.43
Two similar coils, each of 600 turns, are magnetically coupled. It is found that a current of 10 A in either coil produces a flux of 12 mWb in that coil and when this current is reversed in 0·1 s the e.m.f. induced in the other coil is 120 V. Estimate the inductance of each coil, their mutual inductance and the coupling factor.

Example 5.44
Two coils *A* and *B* are so disposed that 80 per cent of the flux produced by one coil cuts the turns of the second coil. Estimate the e.m.f. induced in coil *B*, which has 200 turns when a flux of 5 mWb in *A* is reversed in 0·2 s.

Example 5.45
A primary coil of 1320 turns is wound on a magnetic core together with a secondary coil so that the coupling factor between the two coils is 0·96. How many turns must there be on the secondary so that when a 11 kV a.c. supply is applied to the primary the p.d. across the secondary shall be 3·3 kV?

Chapter 6

Electrostatics and Capacitance

6.1 Electric Field Strength

When an object is charged electrostatically, electric flux is produced and the extent of the electric field is defined as that region in which the influence of the flux may be detected. Unit flux is considered to be associated with unit charge, i.e. one coulomb.

Between any two points in an electric field a force exists and the work done in moving unit charge between the two points is numerically equal to the p.d. between the two points.

Example 6.1
Two points A and B are at positive potentials of 5 V and 15 V respectively. Determine the work done in moving a positive charge of 2 C from A to B.

$$\text{Potential difference between } A \text{ and } B = 15 - 5 \text{ V} = 10 \text{ V}$$
$$\text{Work done} \qquad\qquad\qquad = \text{charge} \times \text{p.d.}$$
$$= 2 \times 10 \text{ J} = 20 \text{ J}$$

The work done in moving the charge is 20 J.

N.B. If the charge had moved from B to A, 20 J of work would have been done by the charge.

Example 6.2
The work done by a positive charge of 2 C, in moving from a point at a positive potential of 6 V to a second point, is 30 J. Determine the potential of the second point.

Let V be the potential difference between the two points. Then

$$V = \frac{\text{work done in J}}{\text{charge in C}}$$
$$= \frac{30}{2} \text{ V} = 15 \text{ V}$$

Since work is being done by the charge, the second point must be at a lower potential than the first point.

$$\text{Potential of second point} = 6 - 15\,V = -9\,V$$

The second point is at a negative potential of 9 V.

Note: Since

$$\text{work done} = \text{charge} \times \text{potential difference}$$
$$= \text{force} \times \text{distance moved}$$

then
$$F \times d = Q \times V$$
$$F = Q \times V/d$$

Hence the force exerted by or on an electric charge is the product of the charge and the potential difference per unit length.

Potential difference per unit length is called the *potential gradient*, denoted by $\mathscr{E}$, and is a measure of the field strength.

Example 6.3
The potential difference between two points 2 mm apart is 240 V. What will be the force exerted on a charge of 0·2 C placed in the field between these two points?

$$\text{Potential gradient between points} = \frac{V}{d} = \frac{240}{2 \times 10^{-3}}\,V/m$$
$$= 120 \times 10^3\,V/m$$
$$\text{Force acting on charge} \qquad = 0{\cdot}2 \times 120 \times 10^3\,N = 24\,kN$$

The force acting on the charge is 24 kN.

Example 6.4
The work done in moving a charge along an electric field between two points, having a potential difference of 50 V, is 125 J. Calculate the value of the charge.

Example 6.5
A charge of 5 C moves between two points, one of which has a positive potential of 10 V. If the work done is 100 J calculate the two possible potentials of the second point.

Example 6.6
The force acting on a charge of 0·5 C, in an electric field, is 20 kN. Calculate the potential gradient of the field. If the field is 2·5 mm in length determine the potential difference between the two points producing the field.

Example 6.7
Given that 1 C is equal to 6.2×10^{18} electrons, show that the energy of 1 electron volt is 0.16×10^{-18} J.

Example 6.8
Two points have a potential difference of 200 V. If the force, acting on a charge of 0·4 C, is 40 kN determine the distance between the two points.

6.2 Electric Flux Density

Since unit flux is associated with unit charge, then unit electric flux density is unit flux per unit area. If the total electric flux is ψ C, then

$$\text{electric flux density, } D = \psi/A \text{ C/m}^2$$

Example 6.9
A charge of 500 μC is given to a metal plate measuring 50 mm × 20 mm. What will be the electric flux density produced?

$$\text{Electric flux } \psi = 500 \times 10^{-6} \text{ C}$$
$$\text{Area of plate } = 50 \times 20 \text{ mm}^2$$
$$= 1000 \times 10^{-6} \text{ m}^2$$
$$D = \frac{\text{electric flux}}{\text{area}}$$
$$= \frac{500 \times 10^{-6}}{10^{-3}} \text{ C/m}^2 = 0.5 \text{ C/m}^2$$

The electric flux density produced will be 0·5 C/m².

Example 6.10
An electric flux density of 1·2 C/m² is required in a field of area 800 mm². Determine the charge which is required.

$$\text{Total flux } \psi = \text{flux density} \times \text{area}$$
$$= 1.2 \times 800 \times 10^{-6} \text{ C}$$
$$= 0.96 \times 10^{-3} \text{ C}$$

The required charge is 0·96 mC.

Example 6.11
A charge of 50 μC is given to the surface of an isolated sphere having a diameter of 100 mm. Calculate the electric flux densities on the surface of the sphere and at a point 50 mm from the sphere.

Example 6.12
A charge of 200 μC produces an electric flux density of 0·4 C/m². What must be the area of the field?

Example 6.13
What charge is required to produce an electric flux density of 0·2 C/m² over an area of 0·8 × 10^{-3} m²?

6.3 Permittivity

When an electric field is produced in air the ratio between the flux density and the field strength is a constant. This constant is called the permittivity of free space, it is denoted by ϵ_0 and is equal to 8·85 × 10^{-12} SI units.

If a potential difference be applied to two conductors separated by some insulating material other than air, then the flux density so produced will be greater than that produced by the same p.d. if the insulator were air.

The ratio between the flux density in the material and the flux density in air is called the relative permittivity of the material. It is denoted by ϵ_r, has no units and its value varies with different materials.

Example 6.14
A charge of 0·02 μC is given to a plate having an area of 0·01 m². Calculate the field strength, in air, between this plate and a similar parallel plate 2·2 mm away. What must be the p.d. between the plates?

$$\text{flux density } D = \frac{Q}{A} = \frac{0\cdot02 \times 10^{-6}}{0\cdot01} \text{ C/m}^2$$

$$= 2\,\mu\text{C/m}^2$$

But $\text{permittivity} = \dfrac{\text{flux density}}{\text{field strength}}$

and so $\text{field strength} = \dfrac{D}{\epsilon_0} = \dfrac{2 \times 10^{-6}}{8\cdot85 \times 10^{-12}} \text{ V/m}$

$$= 0\cdot226 \text{ MV/m}$$

since $\text{field strength} = V/d$

then $\text{p.d. } V = 0\cdot226 \times 10^6 \times 2\cdot2 \times 10^{-3} \text{ V}$

$$= 496 \text{ V}$$

The field strength is 0·226 MV/m and the required p.d. is 496 V.

Example 6.15
In a flat sheet of glass, 5 mm thick and having an area of 0·01 m², the field strength has not to exceed 10 MV/m. If the relative permittivity of the glass

is 6, determine the maximum charge which may be given to the metallic electrodes between which the glass is placed.

Since

$$D/\mathscr{E} = \epsilon_0 \epsilon_r$$

then

$$D = \epsilon_0 \epsilon_r \mathscr{E}$$

Since $\mathscr{E}$ must not exceed 10 MV/m,

maximum flux density $= 8{\cdot}85 \times 10^{-12} \times 6 \times 10 \times 10^6$ C/m

$= 53{\cdot}1 \times 10^{-5}$ C/m $= 0{\cdot}53$ mC/m

maximum charge $= DA = 0{\cdot}53 \times 10^{-3} \times 0{\cdot}01$ C

$= 5{\cdot}3 \times 10^{-6}$ C $= 5{\cdot}3$ μC

The maximum charge which may be given to the electrodes is 5·3 μC.

Example 6.16
A flat sheet of mica, of area 0·05 m² and thickness 0·3 mm, just fills the space between two metal plates. When a p.d. of 250 V d.c. is applied to the plates, the charge given to each plate is 1·8 μC. Estimate the relative permittivity of the mica.

$$\text{Relative permittivity of mica} = \frac{\text{flux density in mica}}{\text{flux density in air}}$$

and

flux density in air $= \epsilon_0 \mathscr{E}$ C/m

Hence for mica $\quad \epsilon_r = D/\epsilon_0 \mathscr{E}$

flux density in mica $= \dfrac{Q}{A} = \dfrac{1{\cdot}8 \times 10^{-6}}{0{\cdot}05}$ C/m²

$= 36 \times 10^{-6}$ C/m²

Potential gradient across mica $= \dfrac{250}{0{\cdot}3 \times 10^{-3}}$ V/m $= 0{\cdot}83 \times 10^6$ V/m

Hence ϵ_r for mica $= \dfrac{36 \times 10^{-6}}{8{\cdot}85 \times 10^{-12} \times 0{\cdot}83 \times 10^6} = 4{\cdot}75$

The relative permittivity of the mica is 4·75.
Note: When an insulating material or dielectric is placed between metallic electrodes a capacitor is formed.

Example 6.17
A potential difference of 240 V d.c. is maintained between two plates separated

by air. If the plates are 1·5 mm apart what is the flux density in the air? How would this flux density be altered if paper 1·5 mm thick were placed between the plates? The paper has a relative permittivity of 2.

Example 6.18
When a p.d. of 250 V d.c. is applied to the plates of a capacitor in which the dielectric is a 3 mm thick sheet of ebonite, the charge is 0·025 μC. If the relative permittivity of the ebonite is 3, determine the effective area of the dielectric.

Example 6.19
A flat sheet of glass, 2·5 mm thick and having a relative permittivity of 6, is placed between two metallic electrodes so that the glass occupies all the space between the electrodes. If the flux density in the glass is 5 μC/m² determine the p.d. between the electrodes.

Example 6.20
A square piece of rubber is 10 mm thick and its sides are 150 mm long. When the rubber is placed between two flat metal plates of the same area and a charge of 0·042 μC is given to each of the plates, the field strength through the rubber is 60 kV/m. Estimate the relative permittivity of the rubber.

6.4 The Parallel-plate Capacitor

Capacitance is that property of a capacitor which enables it to store an electric charge. The fundamental unit of capacitance is the farad (F) and submultiples are the microfarad (μF), the nanofarad (nF) and the picofarad (pF).

$$1\,\mu F = 10^{-6}\,F; \quad 1\,nF = 10^{-9}\,F; \quad 1\,pF = 10^{-12}\,F$$

A capacitor is said to have a capacitance of 1 F when a charge of 1 C produces a p.d. of 1 V between the plates of the capacitor. From this definition

$$Q = CV$$

where Q is the charge in coulombs, C is the capacitance in farads, V is the p.d. in volts.

Example 6.21
In a parallel-plate capacitor the dielectric is of waxed paper having a relative permittivity of 3. The thickness of the dielectric is 0·03 mm and the area of the paper through which the field passes is 6000 mm². Determine from first principles the capacitance of the capacitor.

Let C be the capacitance of the capacitor. Let a charge of Q be given to the capacitor and let the resulting p.d. be V.

Electric flux density in dielectric $= Q/A$

i.e.
$$D = \frac{Q}{6000 \times 10^{-6}} \text{ C/m}^2$$

Potential gradient $\mathscr{E}$
$$= \frac{V}{d} = \frac{V}{0.03 \times 10^{-3}} \text{ V/m}$$

$$D/\mathscr{E} = \epsilon_0 \epsilon_r$$

Hence
$$\frac{Q}{6000 \times 10^{-6}} \times \frac{0.03 \times 10^{-3}}{V} = 8.85 \times 10^{-12} \times 3$$

$$Q/200 \, V = 8.85 \times 10^{-12} \times 3$$

$$Q/V = 8.85 \times 6 \times 10^{-10}$$

But by definition of the farad,

$$Q/V = C$$

and so
$$C = 8.85 \times 6 \times 10^{-10} \text{ F}$$

$$= 53.1 \times 10^{-10} \text{ F}$$

$$= 5.31 \times 10^{-3} \, \mu\text{F} = 5.31 \text{ nF}$$

The capacitance of the capacitor is 5.31 nF (nanofarad).

Note:

Since
$$C = \frac{Q}{V} = \frac{DA}{\mathscr{E}d}$$

But
$$D = \epsilon_0 \epsilon_r \mathscr{E}$$

Hence
$$C = \frac{\epsilon_0 \epsilon_r \mathscr{E}A}{\mathscr{E}d} = \frac{\epsilon_0 \epsilon_r A}{d}$$

N.B. In this expression for the capacitance of a parallel-plate capacitor, A m² is the total area of the dielectric under strain and d is the total thickness. Hence if a capacitor is composed of a number of similar plates connected into two parallel groups then the total area of the dielectric under strain will be the area of one section of the dielectric times the number of such sections.

Example 6.22
The spaces between five flat metal parallel plates forming a capacitor are completely filled by sheets of mica 0.04 mm thick and having a relative permittivity of 4. If the capacitor has a capacitance of 30 nF, determine the area of each sheet of mica under strain.

Since
$$C = \frac{\epsilon_0 \epsilon_r A}{d}$$

then
$$A = \frac{Cd}{\epsilon_0 \epsilon_r} = \frac{30 \times 10^{-9} \times 0.04 \times 10^{-3}}{8.85 \times 10^{-12} \times 4} \text{ m}^2$$

S

$$= 0.0339 \text{ m}^2$$

Since there are 5 plates in the capacitor there must be 4 sheets of mica. Hence

$$\text{area of one sheet} = \frac{0.0339}{4} \text{ m}^2$$

$$= 0.0085 \text{ m}^2 = 8500 \text{ mm}^2$$

The area of each sheet of mica under strain is 8500 mm^2.

Example 6.23

Given that the capacitance of a parallel-plate capacitor is $C = \epsilon_0 \epsilon_r A/d$ F, deduce the unit of the permittivity of free space.

Since $\qquad\qquad\qquad C = \epsilon_0 \epsilon_r A/d$

then $\qquad\qquad\qquad \epsilon_0 = Cd/\epsilon_r A$ SI units

Substituting the fundamental units for the quantities in the right hand expression,

$$\text{SI units for } \epsilon_0 = \frac{F \times m}{m^2} = \frac{F}{m}$$

(as ϵ_r is a ratio it is just a number).

Hence the unit for the permittivity of free space is the farad per metre, i.e. F/m.

Example 6.24

A capacitor is built up of twelve flat parallel sheets of tinfoil, each pair being separated by tissue paper 0.003 mm thick. If the area of the field is 0.015 m^2, estimate the charge required to raise the p.d. between the terminals of the capacitor to 240 V. The tissue paper has a relative permittivity of 3.

$$\begin{array}{ll}
\text{No. of capacitor plates} & = 12 \\
\text{No. of dielectric sections} & = 11 \\
\text{Area of one section under strain} & = 0.015 \text{ m}^2 \\
\text{Total area under strain} & = 0.015 \times 11 \text{ m}^2
\end{array}$$

Let C be the capacitance of the capacitor. Then

$$C = \frac{8.85 \times 10^{-12} \times 3 \times 15 \times 10^{-3} \times 11}{3 \times 10^{-6}} \text{ F}$$

$$= 1.46 \times 10^{-6} \text{ F}$$

$$Q = CV = 1.46 \times 10^{-6} \times 240 \text{ C}$$

$$= 350 \, \mu\text{C}$$

The required charge is 350 μC.

Example 6.25
If a charge of 10 μC be given to a 4 μF capacitor what will be the resulting p.d. between the plates of the capacitor?

Example 6.26
A capacitor with air as the dielectric consists of two parallel plates each measuring 0·15 m × 0·15 m spaced 1·5 mm apart. Obtain from first principles the capacitance of the capacitor.

Example 6.27
A flat sheet of glass, relative permittivity 5 and having an area of 8000 mm², forms the dielectric of a capacitor which has a capacitance of 150 pF. Estimate the thickness of the glass.

Example 6.28
A p.d. of 250 V d.c. is applied to the terminals of a parallel-plate capacitor with mica dielectric 0·03 mm thick. The mica has a relative permittivity of 3. Determine the electric-flux density in the mica.

Example 6.29
The paper dielectric, of a parallel-plate capacitor under working conditions has a potential gradient across the paper of 8 MV/m and electric flux density 0·35 mC/m². Determine the relative permittivity of the paper used.

Example 6.30
A capacitor, having a capacitance of 0·2 μF, is charged by a constant charging current of 0·4 μA for a period of 100 s and is then isolated from the supply and its surroundings. What p.d. will be produced between the terminals of the capacitor? If the dielectric, having a relative permittivity of 3·5, is then withdrawn from the capacitor, what will be the new p.d. across the capacitor? (The charge remains constant.)

Example 6.31
A variable air capacitor consists of 16 fixed flat semicircular vanes, of 35 mm diameter and mounted 0·6 mm apart, and 17 rotating flat vanes of 30 mm diameter and 0·5 mm thick. Determine the maximum capacitance of the capacitor and its capacitance when the rotating vanes have rotated $\pi/6$ rad from the position of maximum capacitance.

6.5 Graphs of Charging Current and Voltage/Time

Since $$Q = CV$$

and $$Q = It$$

then
$$It = CV$$

If a current of I A flows for a period of time denoted by $(t_2 - t_1)$ then the p.d. across the capacitor will change by $(V_2 - V_1)$. Then

$$I(t_2 - t_1) = C(V_2 - V_1)$$
$$I = C\frac{(V_2 - V_1)}{(t_2 - t_1)}$$

Example 6.32
The charging current of a 4 µF capacitor varies with time according to the table given below:

Time intervals in ms	0–20	20–30	30–50	50–80	80–130
Current in mA	10	30	25	0	−20

Draw the current/time curve and draw the curve showing how the capacitor p.d. will vary over the period of 130 s.
   Fig. 6.1 shows the curves drawn to the following scales:

$$\text{Horizontal scale: Time} \quad 1\text{ mm} \equiv 1\text{ ms}$$
$$\text{Vertical scales:} \quad \text{Current } 1\text{ mm} \equiv 1\text{ mA}$$
$$\text{Voltage } 1\text{ mm} \equiv 5\text{ V}$$

Since
$$I(t_2 - t_1) = C(V_2 - V_1)$$

then
$$(V_2 - V_1) = \frac{I}{C}(t_2 - t_1)$$

When $t_1 = 0$ assume that $V_1 = 0$.

$$V_2 - 0 = \frac{10 \times 10^{-3}}{4 \times 10^{-6}}(20 - 0)\,10^{-3} = 50\text{ V}$$

For the second period

$$V_3 - 50 = \frac{30 \times 10^{-3}}{4 \times 10^{-6}}(30 - 20)\,10^{-3} = 75$$
$$V_3 = 75 + 50 = 125\text{ V}$$

Similarly for the third period,

$$V_4 - 125 = \frac{25 \times 10^{-3}}{4 \times 10^{-6}}(50 - 30)\,10^{-3} = 125$$
$$V_4 = 125 + 125 = 250\text{ V}$$

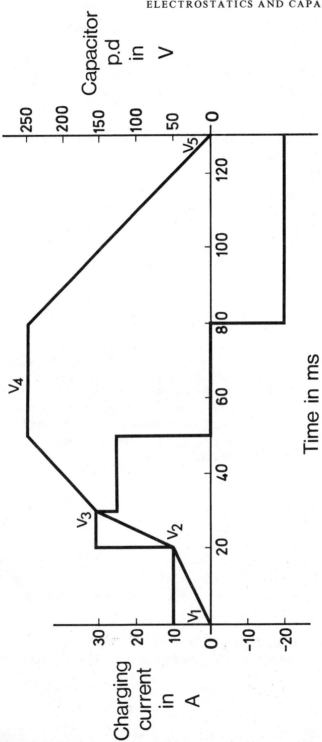

Fig. 6.1 Current and voltage/time curves for Example 6.32 (to scale).

With zero current for the period 50–80 ms, the capacitor p.d. will remain constant at 250 V.

$$V_5 - 250 = \frac{-20 \times 10^{-3}}{4 \times 10^{-6}} (130 - 80) \, 10^{-3} = -250$$

$$V_5 = -250 + 250 = 0 \text{ V}$$

Note: The area under the positive current curve represents the charge given to the capacitor and that under the negative current curve represents the charge taken from the capacitor.

$$\text{Charge given to capacitor} = (10 \times 20) + (30 \times 10) + (25 \times 20) \, \mu C$$
$$= 200 + 300 + 500 \, \mu C = 1000 \, \mu C$$
$$\text{Charge from capacitor} \quad = (20 \times 50) \, \mu C = 1000 \, \mu C$$

Hence the capacitor is fully discharged and this corresponds to the final capacitor p.d. of zero.

In Example 6.32 a circuit of capacitance only was considered and, in order to obtain a curve of charging current as represented in Fig. 6.1, the voltage applied to the circuit would have to follow the voltage curve shown in Fig. 6.1. If, however, a constant p.d. were applied to a circuit of resistance $(R \, \Omega)$ in series with a capacitance $(C \, F)$ then the capacitor p.d./time curve would not be uniform: the rate of growth of the capacitor p.d. would decrease as the capacitor p.d. approached that of the supply.

If the original rate of growth of the capacitor p.d. were to remain constant then the capacitor p.d. would become equal to that of the supply (capacitor fully charged) in a period of time called the *time constant* $(T \, s)$.

For a resistive–capacitive circuit,

$$T = CR$$

Example 6.33
A circuit of resistance 5 MΩ and capacitance 8 μF is connected to a 250 V d.c. supply. Draw the approximate capacitor p.d./time curve and from this curve obtain the value of capacitor p.d. 40 s after the supply is connected to the circuit.

$$\text{Time constant } T = CR = 8 \times 10^{-6} \times 5 \times 10^6 \text{ s}$$
$$= 40 \text{ s}$$

When fully charged the capacitor p.d. will be 250 V. Fig. 6.2(a) shows how the curve may be constructed. Choose suitable voltage and time scales.

$$\text{Horizontal scale: Time} \qquad 1 \text{ mm} \equiv 1 \text{ s}$$
$$\text{Vertical scale:} \quad \text{Capacitor p.d. 1 mm} \equiv 2{\cdot}5 \text{ V}$$

Draw a horizontal line at the final value of the capacitor p.d., i.e. 250 V. Along this line mark a point M at a distance from the vertical equivalent to the time constant of the circuit, i.e. 40 s. Draw a straight line through the origin and the point M.

This line represents the initial rate of growth of the capacitor p.d. Along this line mark point A, which should not be too far from the origin. From A mark off a horizontal distance equal to T and erect a perpendicular to cut the 250 V horizontal at the point M'. Join AM'.

Repeat this same construction for a number of points over a period of time equal to 4T. Draw a smooth curve through the marked points A, A'....

The construction is based on the theory that if, at any instant, the rate of change of current or voltage were thereafter to remain constant, the time taken for the current or voltage to reach their final values would always be equal to the time constant T of the circuit.

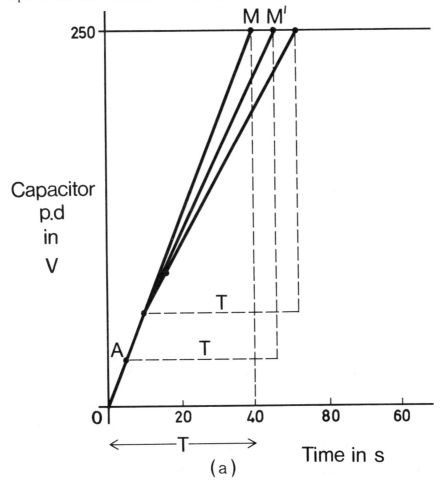

(a)

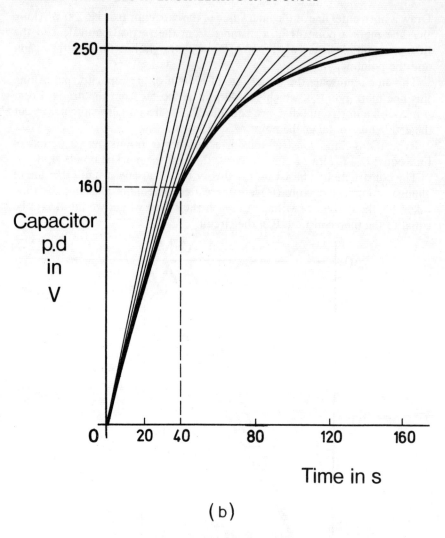

(b)

Fig. 6.2 (a) Construction of capacitor p.d./time curve for Example 6.33.
(b) Capacitor p.d./time curve for Example 6.33 (to scale).

N.B. The voltage or current will always be T away from reaching their final theoretical values.

Fig. 6.2(b) shows the smooth curve drawn to the following scales:

Horizontal scale: Time: 1 mm ≡ 2 s

Vertical scale: Capacitor p.d. 1 mm ≡ 2·5 V

From the curve, when $t = 40$ s, capacitor p.d. = 160 V.

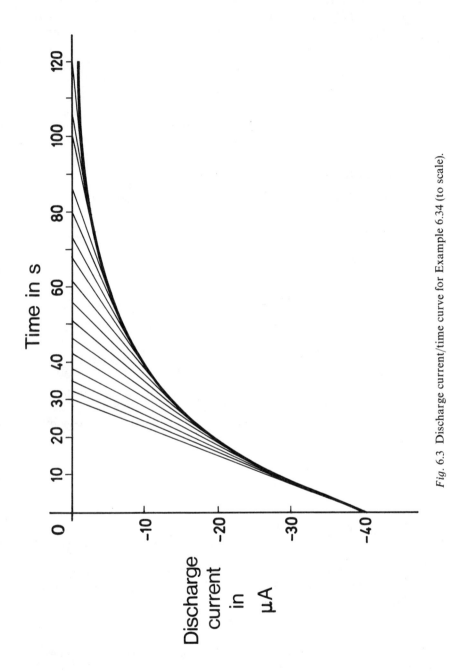

Fig. 6.3 Discharge current/time curve for Example 6.34 (to scale).

Note: This value of capacitor p.d. (160 V) is 64 per cent of the p.d. when the capacitor is fully charged. Theoretically it should be 63·2 per cent of this final p.d. This provides an alternative method of defining the time constant of a circuit.

Example 6.34
A 5 μF capacitor, charged to a p.d. of 240 V, is discharged through a 6 MΩ resistor. Draw the current/time curve for the decay of current, and estimate the value of the current 45 s after the start of the discharge.

Since this example is concerned with a discharge current, the current/time curve should be constructed below the time axis in order to differentiate between the currents for charge and discharge.

$$\text{Time constant for circuit} = 5 \times 10^{-6} \times 6 \times 10^6 \, \text{s} = 30 \, \text{s}$$

The fully charged capacitor will drive maximum current through the resistor and then as the capacitor discharges the current value will decrease.

$$\text{Initial value of current} = -\frac{240}{6 \times 10^6} \, \text{A} = -40 \, \mu\text{A}$$

$$\text{Final value of current} = 0 \, \text{A}$$

Fig. 6.3 shows the current/time curve drawn to the following scales:

$$\text{Horizontal scale: Time} \quad 1 \, \text{mm} \equiv 1 \, \text{s}$$

$$\text{Vertical scale:} \quad \text{Current } 1 \, \text{mm} \equiv 0·5 \, \mu\text{A}$$

The first construction line drawn is between $(t = 0; \, i = -40 \, \mu\text{A})$ and $(t = 30 \, \text{s}; \, i = 0 \, \mu\text{A})$. From the curve, when $t = 45 \, \text{s}, \, i = -8 \, \mu\text{A}$.

N.B. A curve of charging current/time would be the mirror image of the curve shown in Fig. 6.3.

Example 6.35
The charging current of a 10 μF capacitor varies with time according to the following table:

Time (t) in ms	0–20	20–50	50–75	75–100	100–180
Current (i) in mA	30	25	50	0	−30

Draw the current/time curve and the voltage/time curve and from the latter determine (a) the capacitor p.d. when $t = 60$ ms and (b) the charge remaining in the capacitor at the end of the 180 ms period.

Example 6.36
An 8 μF capacitor is charged from a d.c. supply such that the capacitor p.d. varies linearly with time during each interval according to the table given below:

106

Time (t) in ms	0	15	40	60	100
Capacitor p.d. (V) in V	0	100	230	230	0

Draw the current/time curve and hence determine the total charge given to the capacitor during the first 40 ms.

Example 6.37
A capacitive circuit, having a resistance of 0·2 MΩ has a time constant of 1·2 s. What is the capacitance of the circuit?

Example 6.38
Calculate the time constant of a circuit in which a 5 µF capacitor is connected in series with a parallel combination of 3 MΩ and 6 MΩ.

Example 6.39
An 8 µF capacitor, in series with a 6 MΩ resistor, is to be charged from a 240 V d.c. supply. Draw the curves of charging current and capacitor p.d. against time. What will be the volt drop across the resistor, 40 s after the capacitor begins to charge?

Example 6.40
A circuit consists of a 6 µF capacitor in series with a 3 MΩ resistor. What extra resistance is needed, and how must it be connected, to reduce the time constant of the circuit to 10 s?

Example 6.41
Estimate the charge possessed by a 5 µF capacitor, which is connected in series with a 4 MΩ resistor, 20 s after the circuit is connected to a 240 V d.c. supply.

Example 6.42
A 2 µF capacitor is charged to a p.d. of 250 V and then discharged through a 10 MΩ resistor. Draw the curve showing the decay of capacitor p.d./time and hence determine the time taken for the capacitor p.d. to decrease to 100 V.

Example 6.43
In an experiment to determine the time constant of a resistive–capacitive circuit the following results were obtained:

Time in s	0	15	30	45	60	75	90	105	120	135	150
Capacitor p.d. in V	0	89	138	174	196	213	224	231	238	242	244

Draw the capacitor p.d./time curve and hence, by two methods, determine the time constant for the circuit.

107

Chapter 7

Capacitor Connections

7.1 Capacitors in Series

Fig. 7.1 shows two capacitors, C_1 and C_2, in series. As the two plates, which are connected together, will be at the same potential then the equivalent capacitor will consist of the positive plate of C_1 and the negative plate of C_2.

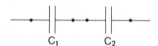

Fig. 7.1 Capacitors in series.

In effect the thickness of the dielectric will have been increased and since capacitance varies inversely with the thickness of the dielectric, the resultant capacitance will be less than that of either C_1 or of C_2.

Example 7.1
Capacitors of 4 µF and 6 µF are connected in series to a 240 V d.c. supply. Determine the equivalent capacitance of the circuit and the p.d. across each capacitor.
   Since the capacitors are connected in series then, by definition of a series circuit, the same current will flow, for the same length of time, in all parts of the circuit. Hence both capacitors will receive the same charge. Let Q coulombs be this charge. Then

$$\text{p.d. across 4 µF capacitor} = \frac{Q}{C} = \frac{Q}{4 \times 10^{-6}} \text{ V}$$

$$\text{p.d. across 6 µF capacitor} = \frac{Q}{6 \times 10^{-6}} \text{ V}$$

But   applied voltage   = sum of separate p.d.s

and so
$$240 = \frac{Q}{4 \times 10^{-6}} + \frac{Q}{6 \times 10^{-6}}$$

$$240 \times 10^{-6} = \frac{6Q + 4Q}{24}$$

$$Q = \frac{240 \times 10^{-6} \times 24}{10} \, \text{C}$$

$$C_T = \frac{Q}{V_T} = \frac{240 \times 10^{-6} \times 2\cdot4}{240} \, \text{F}$$

$$= 2\cdot4 \, \mu\text{F}$$

$$\text{p.d. across 4 } \mu\text{F capacitor} = \frac{240 \times 10^{-6} \times 2\cdot4}{4 \times 10^{-6}} \, V = 144 \, \text{V}$$

$$\text{p.d. across 6 } \mu\text{F capacitor} = \frac{240 \times 10^{-6} \times 2\cdot4}{6 \times 10^{-6}} \, V = 96 \, \text{V}$$

The equivalent capacitance of the circuit is 2·4 μF and the p.d. across the capacitors are 144 V and 96 V.

Note: From Example 7.1 the total capacitance of 4 μF in series with 6 μF is 2·4 μF. But

$$2\cdot4 \times 10^{-6} = \frac{24 \times 10^{-6}}{10} = \frac{4 \times 10^{-6} \times 6 \times 10^{-6}}{4 \times 10^{-6} + 6 \times 10^{-6}}$$

$$C_T = \frac{C_1 \times C_2}{C_1 + C_2}$$

This is the same relationship as for two resistors in parallel,

$$R_T = \frac{R_1 \times R_2}{R_1 + R_2}$$

i.e.
$$\frac{1}{C_T} = \frac{1}{C_1} + \frac{1}{C_2} + \frac{1}{C_3}$$

and for equal capacitances in series,

$$\text{total capacitance} = \frac{\text{capacitance of one capacitor}}{\text{no. of capacitors in series}}$$

Example 7.2

A charge of 1·5 mC is given to a series circuit of three capacitors, having capacitances of 10 μF, 20 μF and 60 μF respectively. Determine each capacitor p.d. and hence obtain the value of the applied voltage. Then show that for a series circuit the reciprocal of the total capacitance is equal to the sum of the reciprocals of the separate capacitances.

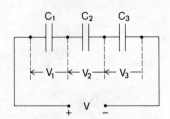

Fig. 7.2 Series circuit for Example 7.2.

Fig. 7.2 shows the circuit and the assumed p.d. for each capacitor. Since

$$Q = C \times V$$

then

$$V = Q/C$$

Hence

$$V_1 = \frac{1 \cdot 5 \times 10^{-3}}{10 \times 10^{-6}} V = 150 \text{ V}$$

$$V_2 = \frac{1 \cdot 5 \times 10^{-3}}{20 \times 10^{-6}} V = 75 \text{ V}$$

$$V_3 = \frac{1 \cdot 5 \times 10^{-3}}{60 \times 10^{-6}} V = 25 \text{ V}$$

Applied voltage $V = V_1 + V_2 + V_3$

$$= 150 + 75 + 25 \text{ V} = 250 \text{ V}$$

Total capacitance, $C_T \qquad = \dfrac{\text{charge}}{\text{total voltage}}$

$$C_T = \frac{1 \cdot 5 \times 10^{-3}}{250} \text{ F}$$

$$= 6 \times 10^{-6} \text{ F} = 6 \, \mu\text{F}$$

$$1/C_T = \tfrac{1}{6}$$

Sum of reciprocals of separate capacitances $= \dfrac{1}{10} + \dfrac{1}{20} + \dfrac{1}{60}$

$$= \frac{6 + 3 + 1}{60}$$

$$= \frac{10}{60} = \frac{1}{6}$$

Hence the reciprocal of the total capacitance is equal to the sum of the reciprocals of the separate capacitances.

Example 7.3
Determine the equivalent capacitance of three capacitors of 3 µF, 8 µF and 24 µF connected in series.

110

Example 7.4
A p.d. of 250 V d.c. is applied to a circuit in which capacitors of 3 µF and 6 µF, are connected in series with an unknown capacitor. If the p.d. across the 3 µF capacitor is 125 V, calculate the charge given to the circuit and the value of the unknown capacitance.

Example 7.5
A circuit consists of two 4 µF capacitors in series and in series with a 5 MΩ resistor. What is the time constant of the circuit?

Example 7.6
Capacitors of 3 µF, 4 µF and 6 µF are connected in series to a 240 V d.c. supply. Calculate the total charge and the p.d. across each capacitor.

Example 7.7
When capacitors of 4 µF, 8 µF and 40 µF are connected in series to an unknown unidirectional supply, the p.d. across the lowest capacitance was measured as 144 V. If a short circuit occurred across the 4 µF capacitor, determine the p.d. across each of the other two capacitors.

7.2 Capacitors in Parallel

Fig. 7.3 shows two capacitors C_1 and C_2 connected in parallel. Since one plate of both of the capacitors is connected to the positive of the supply and the other plate of each capacitor is connected to the negative of the supply, the effect will be as though the two capacitors had been replaced by one capacitor having a plate area equivalent to the sum of the separate areas of the two capacitors.

Since capacitance is directly proportional to the plate area then the parallel connection of capacitors will give an increase in capacitance.

Example 7.8
Three capacitors of 4 µF, 8 µF and 10 µF respectively are connected in parallel to a 250 V d.c. supply. Determine the charge given to each capacitor and the equivalent capacitance of the circuit.

Since the capacitors are connected in parallel then the final p.d. across each capacitor will be equal to the supply voltage, and the total charge taken from the supply will be equal to the sum of the separate charges. Since $Q = CV$,

charge given to the 4 µF capacitor $= 4 \times 10^{-6} \times 250 \, C = 1000 \, \mu C$

charge given to the 8 µF capacitor $= 8 \times 10^{-6} \times 250 \, C = 2000 \, \mu C$

charge given to the 10 µF capacitor $= 10 \times 10^{-6} \times 250 \, C = 2500 \, \mu C$

111

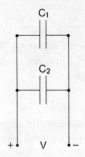

Fig. 7.3 Capacitors in parallel.

total charge taken from supply $= 1000 + 2000 + 2500 \, \mu C$

$$= 5500 \, \mu C$$

But $\quad$ total capacitance $= \dfrac{\text{total charge}}{\text{total voltage}} = \dfrac{5500}{250} \, \mu F$

$$= 22 \, \mu F$$

The equivalent capacitance of the circuit is 22 μF.

Note:

$$\text{Total capacitance } C_T = \frac{\text{total charge}}{\text{total voltage}} = \frac{Q_T}{V}$$

But $\qquad Q_T = Q_1 + Q_2 + Q_3 = C_1 V + C_2 V + C_3 V$

and so $\qquad C_T = \dfrac{V(C_1 + C_2 + C_3)}{V} = C_1 + C_2 + C_3$

Hence the total capacitance of a number of capacitors connected in parallel is equal to the sum of their separate capacitances.

Example 7.9
A 10 μF capacitor, *A*, is connected to a 240 V d.c. supply. When it is fully

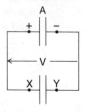

Fig. 7.4 Final combination of capacitors for Example 7.9.

charged it is isolated from the supply and connected to an uncharged 6 μF capacitor, *B*. Determine the final p.d. across the combination.

$$\text{Charge given to } 10\,\mu F \text{ capacitor} = CV = 10 \times 10^{-6} \times 240\,C$$
$$= 2400\,\mu C$$

Fig. 7.4 shows the final combination. As plate *X* of capacitor *B* is connected to the positive terminal of capacitor *A*, then plate *X* will become positively charged and similarly plate *Y* will become negatively charged. Since *A* shares its charge with *B* then the p.d. across *A* will decrease and the p.d. across *B* will increase. This transference of charge will continue until the p.d. across each capacitor is the same. Hence the two capacitors must be connected in parallel. Let *V* be the final p.d. across the combination.

$$\text{Total capacitance} = 10 + 6\,\mu F$$

But   total charge   = total capacitance × final p.d.

and so
$$V = \frac{2400 \times 10^{-6}}{16 \times 10^{-6}}\,V = 150\,V$$

The final p.d. across the combination is 150 V.

Note: The respective charges on the two capacitors will be 1500 μC and 900 μC, giving a total of 2400 μC, which is the original charge in the circuit. When capacitors are charged and then isolated from the supply and their surroundings then the charge in the system must remain constant.

Example 7.10
Capacitors of 6, 12 and 18 μF are connected in parallel to a 200 V d.c. supply. Determine the charge received by each capacitor. Hence determine the total charge given to the system and show that the total capacitance is equal to the sum of the separate capacitances.

Example 7.11
Two capacitors connected in parallel are connected in series with a 4 MΩ resistor. If the time constant of the circuit is 60 s and one of the capacitors has a capacitance of 6 μF determine the value of the second capacitance.

Example 7.12
A capacitor of 8 μF capacitance and with perfect insulation is charged to a p.d. of 240 V and then disconnected from the supply. If it is connected to an uncharged capacitor having a capacitance of 6 μF, estimate the final p.d. of the combination.

Example 7.13
If, in Example 7.12, the 6 μF capacitor had been originally charged to a

T

p.d. of 100 V and then connected in parallel with the charged 8 μF capacitor, what would have been the final p.d. of the combination?

Example 7.14

A capacitor *A* is charged to a p.d. of 250 V and then connected in parallel with a charged capacitor having a capacitance of 8 μF. If the p.d. of the final arrangement is 190 V when a charge of 3·8 mC is present in the system, determine the capacitance of *A* and the potential to which the 8 μF capacitor must have been charged originally.

7.3 Capacitors in Series-Parallel

Example 7.15

Fig. 7.5 shows a system of capacitors connected to a 250 V d.c. supply. Calculate the total capacitance of the system and the p.d. across each capacitor.

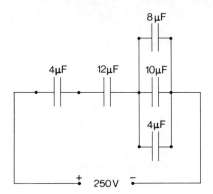

Fig. 7.5 Circuit for Example 7.15.

Consider the two capacitors in series.

$$\text{Equivalent capacitance} = \frac{\text{product}}{\text{sum}} = \frac{4 \times 12}{4 + 12}\mu\text{F}$$

$$= 3 \ \mu\text{F}.$$

Consider the parallel combination.

$$\text{Equivalent capacitance} = \text{sum of the separate capacitances}$$
$$= 8 + 10 + 4\,\mu\text{F} = 22\,\mu\text{F}$$

Fig. 7.6 shows the equivalent circuit.

114

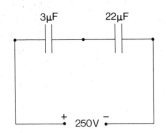

Fig. 7.6 Equivalent circuit for Example 7.15.

Total capacitance of system $= \dfrac{3 \times 22}{3 + 22}\,\mu F = \dfrac{66}{25}\,\mu F$

$= 2{\cdot}64\,\mu F$

Total charge taken by the system = total capacitance × supply p.d.

$$Q_T = 2{\cdot}64 \times 250\,\mu C = 660\,\mu C$$

This will be the charge received by each of the two series capacitors in the original circuit.

p.d. across $4\,\mu F$ capacitor, $V_1 = \dfrac{660 \times 10^{-6}}{4 \times 10^{-6}}\,V = 165\,V$

p.d. across $12\,\mu F$ capacitor, $V_2 = \dfrac{660 \times 10^{-6}}{12 \times 10^{-6}}\,V = 55\,V$

p.d. across parallel combination $= 250 - (165 + 55)\,V$

$= 250 - 220\,V = 30\,V$

Hence each of the three parallel capacitors will be charged to a p.d. of 30 V. The total capacitance of the circuit is $2{\cdot}64\,\mu F$ and the p.d. across each of the separate capacitors are 165 V, 55 V and 30 V.

Note: The p.d. across the parallel combination could also have been found from total charge/total capacitance, i.e.

$$V_3 = \dfrac{660 \times 10^{-6}}{22 \times 10^{-6}}\,V = 30\,V$$

Example 7.16

Figs 7.7 (a), (b), (c) and (d) show four systems in which the capacitances are given in μF. Determine the total capacitance for each of the systems.

Example 7.17

Given four capacitors, each of 4 μF capacitance, determine the total capacitance of all possible arrangements of the four capacitors.

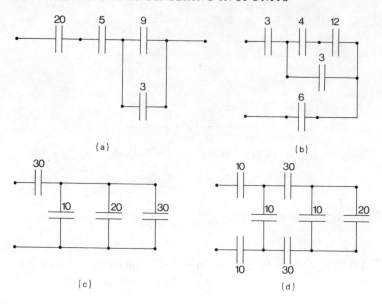

Fig. 7.7 Systems of capacitors for solution in Example 7.16.

Example 7.18

It is required to produce a bank of capacitors giving 25 μF at 1000 V. If the supply of capacitors is limited to those having a capacitance of 10 μF and a p.d. of 250 V, determine how such a bank could be obtained and how many capacitors would have to be used.

Example 7.19

A 5 MΩ resistor is connected in series with a capacitor system consisting of two capacitors in series and in series with a parallel arrangement of a 4 μF and an 8 μF capacitor. If one of the series capacitors has a capacitance of 6 μF, determine the capacitance of the second so that the time constant for the whole circuit may be 12 s.

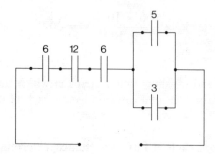

Fig. 7.8. Capacitor circuit for Example 7.20.

Example 7.20

Fig. 7.8 shows a system of capacitors to be connected to a 250 V d.c. supply. If the values given in the diagram are the capacitances in µF, determine the p.d. across each of the capacitors.

7.4 Mixed Dielectrics

Sometimes, in order to control the potential gradient across a dielectric, it is necessary to introduce a second dielectric. In paper-insulated cables a different grade of paper, having a different relative permittivity, is often used as the outer part of the insulation for this purpose.

Example 7.21

Two flat parallel conductors, having a surface area of 0·01 m², are separated by 5 mm of paper insulation consisting of 2 mm of one paper, having a relative permittivity of 3 and 3 mm of a different grade of paper, having a permittivity of 2. Determine the capacitance of the capacitor so formed.

Fig. 7.9(a) shows the arrangement. Since there will be no potential gradient

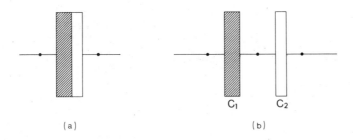

Fig. 7.9 (a) Capacitor with mixed dielectrics.
(b) Equivalent series circuit.

across the junction of the two dielectrics then one surface of each of the dielectrics will be at the same potential.

Fig. 7.9(b) shows two capacitors in series, and in this circuit each capacitor has a plate which is at the same potential as a plate of the other capacitor, i.e. there is no potential gradient between them. Hence the arrangement of mixed dielectrics may be treated as two capacitors in series.

$$\text{Capacitance of first part,} \quad C_1 = \frac{\epsilon_0 \times 3 \times 0\cdot01}{2 \times 10^{-3}} \, F = 15\epsilon_0 \, F$$

$$\text{Capacitance of second part,} \; C_2 = \frac{\epsilon_0 \times 2 \times 0\cdot01}{3 \times 10^{-3}} \, F = \frac{20\epsilon_0}{3} \, F$$

For capacitors in series,

$$C_T = \frac{C_1 \times C_2}{C_1 + C_2} = \frac{15\epsilon_0 \times 20\epsilon_0/3}{15\epsilon_0 + 20\epsilon_0/3}$$

$$= \frac{100\epsilon_0}{65/3} \text{ F} = \frac{300\epsilon_0}{65} \text{ F}$$

$$= \frac{60 \times 8 \cdot 85 \times 10^{-12}}{13} \text{ F} = 41 \text{ pF}$$

The capacitance of the capacitor is 41 pF.

Example 7.22

A parallel plate air capacitor has its plates 10 mm apart. If a dielectric, 10 mm thick and having a relative permittivity of 3, is placed between the plates, by how much must the distance between the plates be increased to restore the capacitance to its original value?

The final arrangement is equivalent to two capacitors in series, one with a dielectric of permittivity 3 and the other with air as the dielectric.

Let A be the area of the dielectric under strain and let d be the amount by which the distance between the plates must be increased.

Original capacitance $= \dfrac{\epsilon_0 A}{10 \times 10^{-3}} \text{ F}$

Capacitance of capacitor with dielectric, $C_1 = \dfrac{\epsilon_0 3A}{10 \times 10^{-3}} \text{ F}$

Capacitance of capacitor with air as dielectric, $C_2 = \dfrac{\epsilon_0 A}{d} \text{ F}$

Total capacitance of combination $= \dfrac{C_1 \times C_2}{C_1 + C_2}$

But $C_1 \times C_2 = \dfrac{\epsilon_0 3A}{0 \cdot 01} \times \dfrac{\epsilon_0 A}{d} \text{ F}^2$

$$= \frac{3\epsilon_0^2 A^2}{0 \cdot 01 \, d} \text{ F}^2$$

and $C_1 + C_2 = \dfrac{3\epsilon_0 A}{0 \cdot 01} + \dfrac{\epsilon_0 A}{d} \text{ F}$

$$= \frac{\epsilon_0 A(3d + 0 \cdot 01)}{0 \cdot 01 \, d} \text{ F}$$

Hence Total capacitance $= \dfrac{3\epsilon_0^2 A^2}{0 \cdot 01 \, d} \times \dfrac{0 \cdot 01 \, d}{\epsilon_0 A(3d + 0 \cdot 01)} \text{ F}$

and so
$$\frac{\epsilon_0 A}{0 \cdot 01} = \frac{3\epsilon_0 A}{3d + 0 \cdot 01}$$
$$3d + 0 \cdot 01 = 0 \cdot 03 \text{ m}$$
$$3d = 0 \cdot 02 \text{ m}$$
$$d = 0 \cdot 02/3 \text{ m} = 6 \cdot 7 \text{ mm}$$

The distance between the plates will have to be increased by 6·7 mm.

Example 7.23

The plates of a parallel-plate capacitor have an area of 0·012 m^2 and are 3 mm apart. If a piece of glass, 3 mm thick and with an area of 9000 mm^2, is placed between the plates, determine the capacitance of the combination. The glass has a relative permittivity of 5.

Since the glass does not occupy the whole area bounded by the plates, the combination consists of two capacitors, one with glass as the dielectric and the other with air as the dielectric. As the p.d. across both parts of the combination is the same it may be considered as being two separate capacitors in parallel.

Capacitance of capacitor with glass as dielectric
$$= \frac{8 \cdot 85 \times 10^{-12} \times 5 \times 9000 \times 10^{-6}}{3 \times 10^{-3}} \text{ F}$$
$$= 132 \cdot 8 \text{ pF}$$

Area of air dielectric $= 12000 - 9000 \text{ mm}^2 = 3000 \text{ mm}^2$

Capacitance of capacitor with air as dielectric
$$= \frac{8 \cdot 85 \times 10^{-12} \times 1 \times 3000 \times 10^{-6}}{3 \times 10^{-3}} \text{ F}$$
$$= 8 \cdot 85 \text{ pF}$$

Hence

Total capacitance of combination $= 132 \cdot 8 + 8 \cdot 85 \text{ pF} = 141 \cdot 65 \text{ pF}$

The total capacitance of the combination is 142 pF.

Example 7.24

A parallel-plate capacitor has its plates 10 mm apart. If a dielectric, 10 mm thick, is placed between the plates, it is found necessary to increase their distance apart by 8 mm in order to restore the capacitance to its original value. Calculate the relative permittivity of the dielectric if its area is equal to that of the plates.

Example 7.25

The parallel plates of an air capacitor are 5 mm apart. If a dielectric having a relative permittivity of 4 is placed between the plates, so as to occupy the whole space between the plates, it is found necessary to increase the distance between the plates in order to restore the capacitance of the capacitor to its original value. Determine the amount by which the distance must be increased.

Example 7.26
When a dielectric, of relative permittivity 3, completely replaces the air between the parallel plates of a capacitor, it is found that the distance between the plates must be increased by 4·5 mm to obtain a capacitor having a capacitance equal to that of the original air capacitor. Calculate the original distance between the plates.

Example 7.27
Determine the capacitance of a parallel-plate capacitor in which the plates have an area of 0·01 m² and the dielectric is compounded 1·5 mm of mica of permittivity 4 and 2 mm of paper of permittivity 2.

Example 7.28
An air capacitor is formed of two flat parallel circular plates, each 0·1 m in diameter and 20 mm apart. Calculate the charge required to raise the p.d. between the plates to 250 V. If a circular piece of glass, of diameter 60 mm and 20 mm thick, is placed centrally between the plates, determine the change which takes place in the p.d. between the plates. The relative permittivity of the glass is 5.

Example 7.29
The flat parallel plates of a capacitor have an area of 0·012 m² and are 2 mm apart. If a 2 mm thick piece of rubber, permittivity 3, having an area of 0·008 m², is placed between the plates what will be the resulting capacitance of the capacitor?

Example 7.30
If in Example 7.29 the rubber had been 1·5 mm thick, what then would have been the capacitance of the combination?

Chapter 8

A.C. Fundamentals

8.1 Instantaneous Value

When a coil is rotated in a magnetic field the value of the induced e.m.f. will vary in magnitude and direction, i.e. it will be an alternating e.m.f.

The instantaneous value of this e.m.f. may be either positive or negative, i.e. it may be in one direction which is assumed to be positive, or in opposition to this direction.

For alternating electrical waveforms, elementary theory is based on the assumption of sine wave form (sinusoidal) or the sum of several sine waves. Let e be the instantaneous value of the e.m.f. which is of sinusoidal form. Then

$$e = E_m \sin \omega t$$

where E_m is the maximum e.m.f. induced in a coil side when the velocity of that coil side is totally perpendicular to the flux.

ω is the angular velocity of the coil in rad/s

t is the time interval, in seconds, which has elapsed since e was zero and increasing in the positive direction.

When the instantaneous value has passed through all its values, ascending and descending, both positive and negative it is said to have completed *one cycle*.

The number of cycles completed in one second is the frequency of the alternating quantity and is measured in Hertz (Hz).

One cycle of induced e.m.f. is completed when a coil side has passed under one pair of poles (p), and electrically, when a coil side has passed under a pair of poles, it is considered to have moved through an angle of 2π (electrical) radians.

Electrical angle = mechanical angle × number of pairs of poles

Example 8.1

A coil is rotated at a speed of 1500 rev/min in the field of a 4-pole machine. Calculate the frequency of the e.m.f. induced in the coil.

Since the speed of rotation is 1500 rev/min, then

$$\text{mechanical angular velocity} = \frac{1500 \times 2\pi}{60} \text{ rad/s} = 50\pi \text{ rad/s}$$

Hence   electrical angular velocity $= 50\pi \times \frac{4}{2} \text{ rad/s} = 100\pi \text{ rad/s}$

and so $\text{frequency } f = \dfrac{100\pi}{2\pi} \text{ Hz} = 50 \text{ Hz}$

The frequency of the induced e.m.f. will be 50 Hz.

Note: $f = \omega/2\pi$ where ω is the angular velocity of the coil measured in rad/s.

Example 8.2
Determine the angular velocity required to generate an e.m.f. having a frequency of 40 Hz.

Since $f = \omega/2\pi$, then
$$\omega = 2\pi f = 2\pi \times 40 \text{ rad/s}$$
$$= 80\pi \text{ or } 251 \cdot 4 \text{ rad/s}$$

To generate an e.m.f. of 40 Hz the angular velocity must be 80π or $251 \cdot 4$ rad/s.

Note: Since $\omega = 2\pi f$, then $e = E_m \sin 2\pi f t$.

Example 8.3
An induced e.m.f. is given by $e = 340 \sin 314t$ V. Determine the time which must elapse, from the instant that the e.m.f. is zero and increasing positively, till the instantaneous e.m.f. is 200 V.

$$e = 340 \sin 314t \text{ V}$$
$$200 = 340 \sin 314t$$
$$\sin 314t = 200/340$$
$$= 0 \cdot 588 = \sin 36°$$

But $314t$ is in radians, and so

$$314t \times \frac{180}{\pi} = 36$$

$$t = \frac{36\pi}{314 \times 180} \text{ s}$$

$$= \frac{\pi}{314 \times 5} \times 10^3 \text{ ms}$$

$$= 2 \text{ ms}$$

The time taken for the induced e.m.f. to increase to 200 V is 2 ms.

Note: The time taken to complete one cycle is called the *periodic time*, usually abbreviated to *period* (*T*).

122

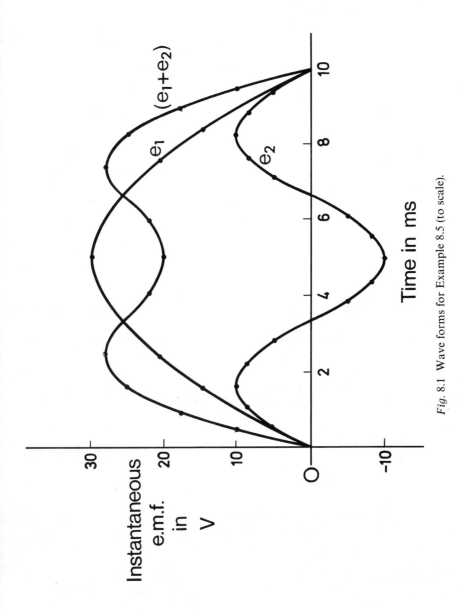

Fig. 8.1 Wave forms for Example 8.5 (to scale).

Example 8.4
Calculate the frequency of an alternating quantity which has a period of 0·1 ms.
Since $T = 1/f$ then

$$f = \frac{1}{T} = \frac{1}{0·1 \times 10^{-3}} \text{Hz}$$
$$= 10\,\text{kHz}$$

The wave form with a period of 0·1 ms has a frequency of 10 kHz.

Example 8.5
An alternating voltage is composed of two sinusoidal voltages of the form $e_1 = 30 \sin 100\pi t$ and $e_2 = 10 \sin 300\pi t$. Draw the wave forms for a time interval of 10 ms and hence draw the wave of the compounded voltage over this same period. What is the maximum value of this voltage?
   Fig. 8.1 shows the wave forms over a period of 10 ms drawn to the following scales:

Horizontal scale: Time $\qquad\qquad$ 1 mm $\equiv$ 0·1 ms
Vertical scale: Instantaneous values of voltage 1 mm $\equiv$ 0·5 V

The compounded wave form is obtained by adding, algebraically, the instantaneous values of e_1 and e_2.
   From the curves the maximum value of the compounded voltage is 28 V.

Example 8.6
At what speed must the rotor of a 6-pole alternator be driven so that it will generate an e.m.f. having a frequency of 50 Hz?

Example 8.7
A coil is rotated at 1000 rev/min in the field due to a pair of electromagnets. Determine the frequency of the e.m.f. induced in the coil.

Example 8.8
What will be the frequency of the e.m.f. induced in a coil having an angular velocity of (a) 90π rad/s (b) 1572 rad/s?

Example 8.9
What must be the relative angular velocity between the magnetic flux and a coil so that the e.m.f. induced in the coil will have a frequency of (a) 16·7 Hz (b) 1 kHz?

Example 8.10
Calculate the periodic times for the frequencies given in Example 8.9.

Example 8.11
An alternating e.m.f. is represented by $e = 283 \sin 314t$ V. Determine the value of the e.m.f. 2 ms after it has passed through its positive maximum value. After what time interval from the start of the cycle will its value be 120 V?

Example 8.12
Draw curves representing $i_1 = 10 \sin 314t$ A and $i_2 = 3 \sin 1570t$ A. Hence obtain the curve for $i_1 - i_2$. What is the value of $i_1 - i_2$, 1·5 ms after i_1 is zero and increasing positively. After what time intervals will $i_1 - i_2$ be equal to $+5$ A?

8.2 Average Value

The average value is the mean value of all the instantaneous values. For symmetrical alternating wave forms the average value taken over a complete cycle is zero. Therefore, when the average value of an alternating wave form is referred to it means the average value over one half of the wave taken between successive zero values.

For a sine wave the average value is $0.636 \times$ the maximum value. The average value may be obtained from the wave form by the use of (a) the mid-ordinate rule or (b) Simpson's rule.

When calculating the value of an induced e.m.f. it is often necessary to use an average rate of change of flux. This then gives an average value for the induced e.m.f.

$$E_{av} = T \times \text{average rate of change of flux}$$

where T is the number of turns cut by the flux.

Example 8.13
Using Simpson's rule, determine the average value of an alternating current of the form $i = 10 \sin 314t$ A.

Fig. 8.2 shows half of the wave form drawn to the following scales:

Horizontal scale: Angular displacement (ωt) 1 mm $\equiv 3°$

Vertical scale: Current (i)   1 mm $\equiv 0·1$ A

To use Simpson's rule the horizontal axis has to be divided into an even number of equal sections, i.e. in Fig. 8.2, at distances equivalent to 7·5°.

Since the half wave is symmetrical about the maximum point then only the first quarter of the cycle need be considered.

Number of ordinate	1	2	3	4	5	6	7	8	9	10	11	12	13
Value of ordinate in A	0	1·3	2·6	3·8	5	6·1	7·1	7·9	8·7	9·2	9·7	9·9	10

125

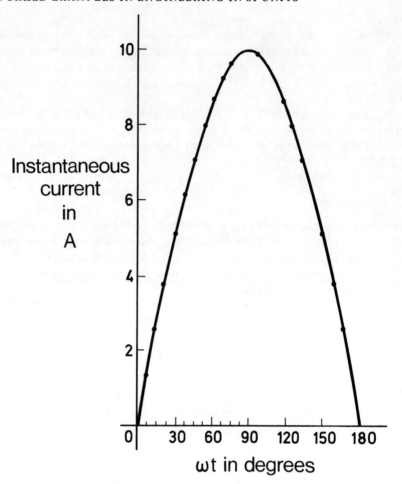

Fig. 8.2 Half wave for the determination of the average value of a sine wave (to scale).

By Simpson's rule,

$$\text{Area} = \frac{\text{time interval}}{3}\left[\text{1st ordinate} + \text{last ordinate} + 4(\text{sum of even ordinates})\right.$$

$$\left. + 2(\text{sum of odd ordinates})\right]$$

The sum of the odd ordinates does not include the first nor the last ordinate.

Sum of even ordinates $= 1\cdot3 + 3\cdot8 + 6\cdot1 + 7\cdot9 + 9\cdot2 + 9\cdot9$ A $= 38\cdot2$ A

Sum of odd ordinates $= 2\cdot6 + 5 + 7\cdot1 + 8\cdot7 + 9\cdot7$ A $= 33\cdot1$ A

(excluding first and last).

126

$$\text{Area} = \frac{7\cdot5}{3}\left[0+10+(4\times38\cdot2)+(2\times33\cdot1)\right] \text{ degree amperes}$$

$$= 2\cdot5[10+152\cdot8+66\cdot2] \text{ degree amperes}$$

$$= 2\cdot5\times229 \text{ degree amperes}$$

$$\text{Average value} = \frac{\text{area}}{\text{length of wave form}}$$

$$= \frac{2\cdot5\times229}{90} \text{ A} = 6\cdot36 \text{ A}$$

The average value of the current is 6·36 A.

Note: If the mid-ordinate rule is used to determine the average value the even ordinates may be used as the mid ordinates.

$$\text{Average value} = \frac{38\cdot2}{6} \text{ A} = 6\cdot37 \text{ A}$$

Example 8.14
A coil of 20 turns is mounted at right angles to a uniform magnetic flux of 5 mWb. When the flux increases for 2 ms the average e.m.f. induced in the coil is 50 V. Determine the value of the flux at the end of the 2 ms.

$$\text{Average e.m.f.} = T\times\text{average rate of change of flux}$$

$$= T\times\frac{\phi_2-\phi_1}{t_2-t_1}$$

Hence
$$50 = 20\times\frac{\phi_2-5\times10^{-3}}{2\times10^{-3}}$$

$$5\times10^{-3} = \phi_2-5\times10^{-3}$$

$$\phi_2 = 5\times10^{-3}+5\times10^{-3} = 10\times10^{-3}$$

The final flux cutting the coil will be 10 mWb.

Example 8.15
Using Simpson's rule determine the average value of the curve obtained in Example 8.12 for the curve of i_1-i_2.

Example 8.16
Using the mid-ordinate rule estimate the average value of an alternating wave form which is semi circular in form and which has a maximum value of 200 V.

Example 8.17
The instantaneous values of an alternating voltage measured at equal time

intervals are as follows:

$$0 \quad 60 \quad 80 \quad 87 \quad 92 \quad 94 \quad 90 \quad 85 \quad 70 \quad 40 \quad 0.$$

Determine the average value of the alternating voltage if these ordinates are joined by straight lines.

Example 8.18
A flux is represented by $\phi = 8 \times 10^{-3} \sin 314t$ Wb. What is the average rate of change of this flux?

Example 8.19
A coil of 15 turns is mounted at right angles to a magnetic flux of 10 mWb which reverses its direction every 5 ms. Estimate the average e.m.f. induced in the coil.

Example 8.20
A coil of 20 turns is rotated at a speed of 1500 rev/min in the field of a machine having 4 poles. If each of the poles produces a magnetic flux of 20 mWb, determine the average e.m.f. induced in the coil.

Example 8.21
When a coil of 25 turns is rotated in the field of a 6-pole machine the generated e.m.f. has a frequency of 50 Hz. If the flux/pole is 10 mWb, determine the value of the induced e.m.f.

8.3 The Root Mean Square Value

The root mean square value (r.m.s.) is that value of a direct current or voltage which produces, in the same circuit and in the same time, the same heating effect as the alternating current or voltage.

For a sine wave,

$$\text{root mean square value} = \text{maximum value}/\sqrt{2}$$

i.e.

$$I = I_m/\sqrt{2} = 0 \cdot 707 I_m$$

or

$$V = 0 \cdot 707 V_m$$

The r.m.s. value is that value indicated by a.c. measuring instruments.

Example 8.22
An alternating voltage is of the form $v = 340 \sin 314t$ V and is connected to a 15Ω resistor. Determine the readings on an ammeter and a voltmeter connected in the circuit.

Since the instruments indicate r.m.s. values and since the supply is sinu-

128

soidal then the voltmeter will read $0.707V_m$

$$\text{Voltmeter reading } V = 0.707 \times 340 \text{ V} = 240 \text{ V}$$

But $$I = V/R$$

and so $$\text{Ammeter reading} = 240/15 \text{ A} = 16 \text{ A}$$

The voltmeter will read 240 V and the ammeter 16 A.

Example 8.23
The ordinates of half an alternating voltage wave, measured at equal time intervals, are given below:

0	10	18	24	27	28	27	25	21	20
21	25	27	28	27	24	18	10	0	

If these points are joined by straight lines, determine the r.m.s. value of the wave form.

Fig. 8.3 shows the half cycle of the wave form. The root mean square value is the value of the square root of the average of the voltage squared wave. The average of the voltage squared wave may be calculated by using the mid ordinate rule. The mid ordinates may be determined from the curve or from the ordinates given.

$$\text{Sum of (mid ordinates)}^2 = 5^2 + 14^2 + 21^2 + 25.5^2 + 27.5^2 + 27.5^2 + 26^2$$
$$+ 23^2 + 20.5^2$$

(Since the wave is symmetrical only half of the wave need be considered.)

$$\text{Sum of (mid ordinates)}^2 = 25 + 196 + 441 + 650 + 756 + 756 + 676 + 529$$
$$+ 420$$

$$= 4449 \text{ V}^2$$

$$\text{Number of mid ordinates} = 9$$

$$\text{Mean square value} = \frac{4449}{9} \text{ V}^2 = 494 \text{ V}^2$$

Hence $$\text{Root mean square value} = \sqrt{494} = 22.2 \text{ V}$$

The r.m.s. value of the given wave form is 22.2 V.

Example 8.24
The e.m.f.s induced in two coils are represented by $e_1 = 340 \sin 314t$ V and $e_2 = 340 \sin(314t - \frac{2}{3}\pi)$ V. If the two coils are connected in series determine the value of the sum of the two e.m.f.s.

Since the two e.m.f.s are of the same frequency $(314/2\pi)$ they may be represented on a *phasor diagram*. A phasor is a straight line which represents a quantity, in sense and magnitude, when the sense and magnitude of that quantity do not vary with time, i.e. r.m.s. values.

129

U

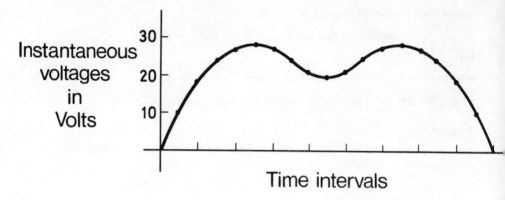

Fig. 8.3 Half cycle of wave form in Example 8.23 (to scale).

Since the e.m.f.s have equal maximum values, 340 V and since they are of sine wave form then

$$\text{r.m.s. value of either e.m.f.} = 0\cdot707 \times 340 \text{ V} = 240 \text{ V}$$

Fig. 8.4 shows the two e.m.f.s drawn to a scale of 1 mm ≡ 4 V. The resultant of the two e.m.f.s may be obtained by the parallelogram rule. From the phasor diagram.

$$\text{length of resultant} = 60 \text{ mm}$$
$$\text{total of } E_1 + E_2 \quad = 60 \times 4 \text{ V} = 240 \text{ V}$$

The r.m.s. value of the sum of the two e.m.f.s is 240 V.

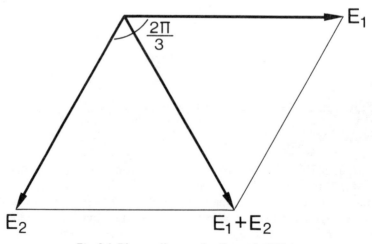

Fig. 8.4 Phasor diagram for Example 8.24.

130

Example 8.25
A current given by $i = 10 \sin 300t$ A is maintained in a resistance of 6 Ω. Determine the expression for the voltage across the resistance and the corresponding reading on an a.c. voltmeter.

Example 8.26
A sinusoidal voltage has an average value of 218 V. Determine the maximum value of this voltage and hence calculate the r.m.s. value.

Example 8.27
A current has a triangular wave form. In 6 ms the current increases linearly from zero to 15 A and then decreases linearly to zero in 4 ms. Determine the r.m.s. value of the wave form.

Example 8.28
Draw the curves of $i_1 = 5 \sin(314t + 30°)$ A and $i_2 = 3 \sin(314t - 45°)$ A. Hence obtain the r.m.s. value for the curve, $i_1 + i_2$.

Example 8.29
Consider the r.m.s. values of a rectangular wave, a sine wave and a triangular wave in terms of their equal maximum values, and hence describe curves having r.m.s. values of 0·85 and 0·5× the maximum value.

Example 8.30
In a network a conductor *AB* is connected to two branches *BC* and *BD*. If the current in *BC* is represented by $i_1 = 3·4 \sin(314t + 30°)$ A and $i_2 = 6·8 \sin(314t - 37°)$ A, represents the current in *BD*, draw the phasor diagram and hence determine the current in *AB*.

8.4 Form Factor (K_f)

The ratio between the r.m.s. value and the half-wave average value of a wave form gives an indication of the shape of the wave. Hence this ratio is called the form factor and is denoted by K_f.

$$K_f = \frac{\text{r.m.s. value}}{\text{average value}}$$

For a pure sine wave,

$$\text{r.m.s. value} = 0·707 \times \text{maximum value}$$
$$\text{average value} = 0·636 \times \text{maximum value}$$

Hence
$$K_f = \frac{0·707}{0·636} = 1·11$$

A wave form with a form factor less than 1·11; a rectangular wave for which $K_f = 1$ indicates a flat-topped wave, whereas a wave with a form factor of more than 1·11 indicates a peaky wave, e.g. a symmetrical triangular wave form has a form factor of 1·157.

In calculations which involve an average rate of change, the form factor is useful in converting from the calculated average value to the r.m.s. value as measured by a.c. instruments.

$$K_f = \frac{\text{r.m.s. value}}{\text{average value}}$$

and so $\qquad$ r.m.s. value $= K_f \times$ average value

Example 8.31
An alternating flux given by $\phi = 0\cdot01 \sin 400t$ Wb cuts a coil of 15 turns. If the coil is connected to a $10\,\Omega$ resistor, calculate the reading on an a.c. ammeter connected in series with the resistor.

$$\text{Change in flux/cycle} = 4\Phi_m \text{ Wb} = 4 \times 0\cdot01 \text{ Wb}$$

$$\text{Average rate of change of flux} = 0\cdot04 \times \frac{400}{2\pi} \text{ Wb/s}$$

$$\text{Average e.m.f. induced in coil} = 15 \times \frac{16}{2\pi} \text{ V}$$

Hence $\quad$ average current $\quad = \dfrac{120}{\pi} \times \dfrac{1}{10} \text{ A} = \dfrac{12}{\pi} \text{ A}$

Since the flux is of sine wave form then the induced e.m.f. and the current will be of the sine wave form. Therefore

$$K_f = 1\cdot11$$

$$\text{ammeter reading} = 1\cdot11 \times \frac{12}{\pi} \text{ A}$$

$$= 4.25 \text{ A}$$

The ammeter would read 4·25 A.

Example 8.32
Determine the form factor for the wave form given in Example 8.23.
$\quad$ From Fig. 8.3 and the list of ordinates given,

$\quad$ Sum of mid ordinates $= 5+14+21+25\cdot5+27\cdot5+27\cdot5+26+23+20\cdot5$
(over half of the wave). $\qquad = 190$ V

$\quad$ Number of ordinates $= 9$

Hence   average value $= \dfrac{190}{9}$ V $= 21 \cdot 1$ V

From Example 8·23,

$$\text{r.m.s. value} = 22 \cdot 2 \text{ V}$$

Hence   form factor $= \dfrac{22 \cdot 2}{21 \cdot 1} = 1 \cdot 052$

The form factor of the wave given in Example 8.23 is 1·052.

Example 8.33
An a.c. ammeter in a circuit reads 4 A. If the wave form of the current is sinusoidal estimate the maximum and average values of the current wave form.

Example 8.34
In a transformer a magnetic flux density given by $b = 1 \cdot 0 \sin 314t$ T cuts a winding of 200 turns. If the effective area of the magnetic circuit is $0 \cdot 01$ m^2 calculate the reading on an a.c. voltmeter connected across the winding.

Example 8.35
Determine the form factor of the triangular wave form described in Example 8.27.

Example 8.36
An alternating voltage has a period of 20 ms and varies over half a cycle in the following manner. For 3 ms it increases linearly from zero to 110 V and remains constant at 110 V for 5 ms. The voltage then decreases linearly to zero in 2 ms. Obtain the form factor of this wave.

Chapter 9

Circuits Connected to an A.C. Supply

When dealing with the effect of different types of circuit connected to an a.c. supply it is easier in some cases to consider the effect of a purely theoretical circuit before considering the practical circuit. For instance, it is not possible to wind an inductive coil without that coil having resistance, but it is easier to work from a consideration of circuits of pure resistance or pure inductance to a circuit possessing both resistance and inductance.

9.1 Circuit of Pure Resistance

Fig. 9.1 shows a circuit of pure resistance $R\,\Omega$ connected to an a.c. voltage of the form $v = V_m \sin \omega t$.

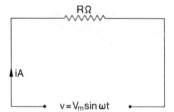

Fig. 9.1 Circuit of pure resistance connected to an a.c. supply.

The current i at any instant, must obey Ohm's law,

$$i = v/R$$

Substituting for v,

$$i = V_m/R \sin \omega t = I_m \sin \omega t$$

Hence the current is of the same form as the applied voltage—both sine waves

134

—has the same frequency ($\omega = 2\pi f$ rad/s) and passes through the same relevant points at the same instant as the voltage, i.e. when $v = 0$, $i = 0$,

$$v = V_m, \qquad i = I_m$$

The current and voltage are said to be *in phase*. Fig. 9.2(a) shows the current

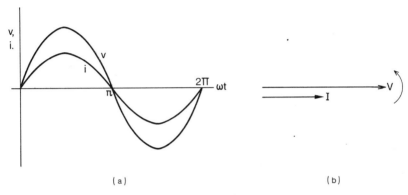

(a)   (b)

Fig. 9.2 (a) Current and voltage wave forms for circuit in Fig. 9.1.
(b) Phasor diagram for circuit in Fig. 9.1.

and voltage waves and Fig. 9.2(b) shows the same relationship by means of phasors.

Example 9.1
A voltage of the form $v = 340 \sin 314t$ V is applied to a 10 Ω resistor. What is the reading of an ammeter in the circuit and what is the form of the current?
Fig. 9.3 shows the circuit diagram. Since

$$v = 340 \sin 314t \text{ V}$$

then $\qquad\qquad V_m = 340 \text{ V}$

and so $\qquad\qquad I_m = \dfrac{340}{10} \text{ A} = 34 \text{ A}$

$$i = 34 \sin 314t \text{ A}$$

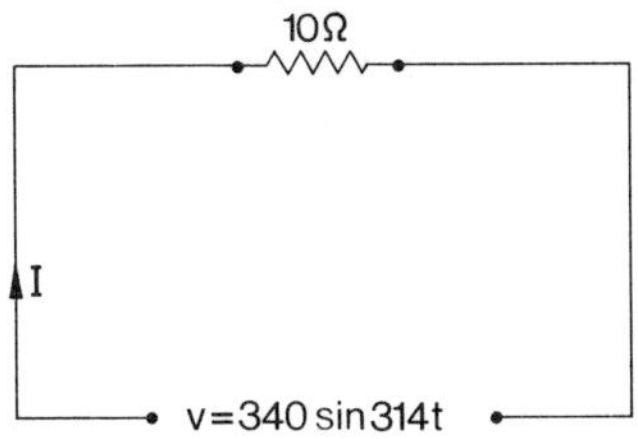

Fig. 9.3 Circuit diagram for Example 9.1.

But the ammeter will read the r.m.s. value of the current, I A, and since both the voltage and the current are of sine wave form,

$$I = 0{\cdot}707 \times I_m$$
$$= 0{\cdot}707 \times 34 \text{ A} = 24.04 \text{ A}$$

The ammeter will read 24 A and the current will be of the form $i = 34 \sin 314t$ A.

Example 9.2
A 20 Ω resistor is connected to an a.c. supply having a frequency of 50 Hz. If an ammeter in the circuit reads 5.3 A, determine the equation for the instantaneous value of voltage across the resistor.

Example 9.3
An alternating voltage of the form $v = 707 \sin 314t$ V is applied to a circuit of pure resistance. If an ammeter in the circuit reads 25 A, calculate the resistance of the circuit.

Example 9.4
When a resistance of 40 Ω is connected to a sinusoidal alternating voltage, having a frequency of 50 Hz, a voltmeter connected across the resistor reads 240 V. Determine expressions for the instantaneous values of the current and the voltage.

Example 9.5
The calibration constant for a cathode ray oscilloscope is 0.2 mm/V. When an alternating voltage is applied to the Y plates and a suitable time base to the X plates, the distance between the peaks of the sine wave trace is measured as 113 mm. What current will flow when this same alternating voltage is applied to a circuit of 10 Ω resistance?

9.2 Circuit of Pure Inductance

Fig. 9.4 shows a circuit of pure inductance, L, in which a current of the form $i = I_m \sin \omega t$ is considered to flow. Since there is no resistance in the circuit then at any instant the applied voltage need only be equal and opposite to the self-induced e.m.f., i.e.

$$v = e_L$$

But $\qquad e_L = L \times \text{rate of change of current}$

and so $\qquad v = L \times \text{rate of change of current}$

136

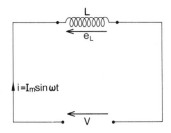

Fig. 9.4 Circuit of pure inductance connected to an a.c. supply.

The current changes from zero to I_m in one quarter of a cycle, and so

$$\text{average rate of change of current} = \frac{I_m}{1/4f} = 4fI_m$$

But $$f = \omega/2\pi$$

Hence average rate of change of current $= 4\omega I_m/2\pi$

average value of voltage, V_{av} $= L \times 2\omega I_m/\pi$

But $$V_m/V_{av} = \pi/2$$

Hence $$V_m = \pi/2 \times V_{av}$$

$$= \frac{\pi}{2} \times \frac{2\omega L I_m}{\pi}$$

$$\frac{V_m}{I_m} = \frac{V}{I} = \omega L$$

This quantity ωL is called the *inductive reactance*: it is denoted by X_L and is measured in ohms.

Fig. 9.5 shows the wave form for the current and from this curve, since the e.m.f. of self inductance is proportional to the rate of change of current, the curves for the induced e.m.f. and the applied voltage may be deduced.

When $\omega t = 0$ the rate of change of current is a positive maximum. Therefore e_L is a positive maximum v is a positive maximum.

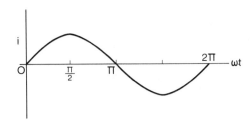

Fig. 9.5 Current wave form for circuit in Fig. 9.4.

137

When $\omega t = \pi/2$ the rate of change of current is zero, e_L is zero and v is zero.

When $\omega t = \pi$ the rate of change of current is a negative maximum, e_L is a negative maximum and v is a negative maximum.

This may be continued for the second half-cycle of current and, assuming that since the current is of sine wave form the voltage will also be sinusoidal, the curves for these voltages may be drawn as in Fig. 9.6(a). Comparing the voltage curve in Fig. 9.6(a) to the current curve in Fig. 9.5, it will be seen that at all instants the voltage passes through a particular value $\pi/2$ rad ahead of the current.

Fig. 9.6(b) shows the phasor diagram representing this condition. (N.B.

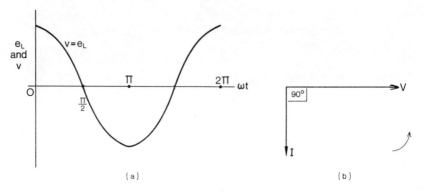

(a)   (b)

Fig. 9.6 (a) Derived voltage curve for circuit in Fig. 9.4.
(b) Phasor diagram for circuit of pure inductance.

The anticlockwise direction of rotation is always assumed as the positive direction of rotation.) The current is said to lag on the voltage by $\pi/2$ rad or 90°.

Example 9.6
A pure inductance of 80 mH is connected to a 250 V, 50 Hz sinusoidal supply. Determine the equation for the current wave produced.

Fig. 9.7 shows the circuit diagram.

Since $$X_L = \omega L$$
and $$\omega = 2\pi f$$
then $$X_L = 2\pi f L$$
$$= 2\pi \times 50 \times 80 \times 10^{-3} \ \Omega$$
$$= 8\pi \ \Omega = 25\cdot1 \ \Omega$$

But $$I = \frac{V}{X_L} = \frac{250}{25\cdot1} \ A$$

$$= 10 \text{ A},$$

and $$I_m = \sqrt{2} \times I \qquad \text{(sine wave form)}$$
$$= \sqrt{2} \times 10 \text{ A} = 14{\cdot}14 \text{ A}$$

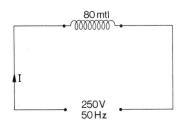

Fig. 9.7 Circuit diagram for Example 9.6.

Since the current will be of the same frequency as the supply voltage, i.e. 50 Hz, then

$$\omega = 2\pi \times 50 \text{ rad/s} = 314 \text{ rad/s}$$

Since the circuit is of pure inductance the current will lag on the voltage by $\pi/2$ rad, then

$$i = 14{\cdot}1 \sin(314t - \pi/2) \text{ A}$$

which is the equation for the current wave form.

Example 9.7
In a coil of 200 turns, connected to a d.c. supply, the change of flux with current is 0·5 mWb/A. If, on a 240 V, 50 Hz sinusoidal supply, this flux/current relationship is constant, determine the current in the coil and the maximum value of flux.

$$\text{Inductance, } L = \frac{\phi_2 - \phi_1}{i_2 - i_1} \times T$$
$$= 0{\cdot}5 \times 10^{-3} \times 200 \text{ H} = 0{\cdot}1 \text{ H}$$

Hence $$\text{reactance, } X_L = 2\pi f L = 2\pi \times 50 \times 0{\cdot}1 \Omega$$
$$= 31{\cdot}4 \ \Omega$$

But $$I = \frac{V}{X_L} = \frac{240}{31{\cdot}4} \text{ A}$$
$$= 7{\cdot}65 \text{ A}$$

Maximum flux will occur when the current is a maximum.

$$I_m = \sqrt{2} \times 7{\cdot}65 \text{ A} \qquad \text{(sine wave)}$$
$$= 10{\cdot}82 \text{ A}$$

Hence $\Phi_m = 0.5 \times 10.82 \text{ mWb} = 5.41 \text{ mWb}$

The current in the coil will be 7·65 A and the maximum flux produced will be 5.41 mWb.

Example 9.8
When a sinusoidal supply of 240 V, 50 Hz is applied to a pure inductance, a current of 8 A flows. What is the value of the inductance?

Example 9.9
A current represented by $i = 3.54 \sin 314t$ A flows in a pure inductance of 0·6 H. Determine the equation for the voltage across the inductor.

Example 9.10
A 250 V, 50 Hz supply is applied to a coil of pure inductance and a current of 8 A is produced. What current would the same coil take if the frequency of the supply were increased to 60 Hz, the voltage remaining constant?

Example 9.11
An inductor is connected to a sinusoidal supply of 240 V but of unknown frequency. If the inductor has an inductance of 40 mH and the current is 21·3 A, determine the frequency of the supply.

Example 9.12
An air-cored coil, having a magnetic length of 150 mm and a diameter of 50 mm, takes a current of 16 A from a 50 V, 50 Hz supply. Neglecting the resistance of the winding, estimate the number of turns in the coil.

9.3 Circuit of Pure Capacitance

Fig. 9.8 shows the circuit of pure capacitance connected to a supply of the form $v = V_m \sin \omega t$.
  At any instant,
$$q = Cv$$
$$i(t_2 - t_1) = C(v_2 - v_1)$$
$$i = C \times \frac{v_2 - v_1}{t_2 - t_1}$$
$$= C \times \text{rate of change of voltage}$$

Fig. 9.9 shows the voltage curve over one cycle.
  When $\omega t = 0$, $v = 0$; but rate of change of voltage is a positive maximum and current is a positive maximum
  When $\omega t = \pi/2$, $v = V_m$; rate of change of voltage is zero and current is zero.

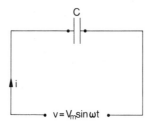

Fig. 9.8 Circuit of pure capacitance connected to an a.c. supply.

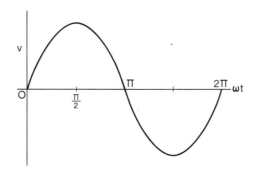

Fig. 9.9 Wave form of voltage applied to the circuit in Fig. 9.8.

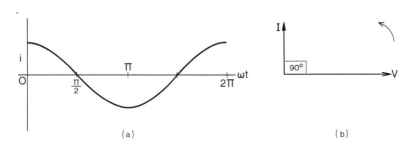

Fig. 9.10 (a) Derived current wave form.
(b) Phasor diagram for circuit of pure capacitance.

When $\omega t = \pi$, $v = 0$ but rate of change of voltage is a negative maximum and current is a negative maximum.

Similarly for the second half of the cycle.

Fig. 9.10(a) shows the current curve deduced as above. By comparison with the voltage curve it is seen that the current leads on the voltage by $\pi/2$ or 90°.

Fig. 9.10(b) shows the phasor diagram for the circuit. The change in voltage in one quarter of a cycle is V_m, and so the average rate of change is $4V_m f$. But $f = \omega/2\pi$, hence

$$\text{average rate of change of voltage} = \frac{4\omega V_m}{2\pi} = \frac{2\omega V_m}{\pi}$$

Since the voltage is of sine wave form,

$$\text{maximum rate of change of voltage} = \frac{\pi}{2} \times \text{average rate of change}$$

$$= \frac{\pi}{2} \times \frac{2\omega V_m}{\pi} = \omega V_m$$

But $\qquad I_m = C \times$ maximum rate of change of voltage

$$= C\omega V_m$$

Hence $\qquad \dfrac{V_m}{I_m} = \dfrac{V}{I} = \dfrac{1}{\omega C}$

$1/\omega C$ is the capacitive reactance, X_C, and is measured in ohms.

$$X_C = \frac{1}{\omega C} = \frac{1}{2\pi f C}\,\Omega = \frac{10^6}{2\pi f C}\,\Omega$$

where C is in microfarads.

Example 9.13

A 40 μF capacitor carries a current of the form $i = 4\cdot25 \sin 314t$ A. Determine the reading on a voltmeter connected across the capacitor and the equation for the supply voltage.

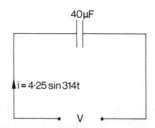

Fig. 9.11 Circuit diagram for Example 9.13.

Fig. 9.11 shows the circuit diagram. From the current equation,

$$\omega = 314 \text{ rad/s}$$

$$X_c = \frac{10^6}{314 \times 40}\,\Omega = 79\cdot5\,\Omega$$

$$V_m = I_m \times X_C = 4\cdot25 \times 79\cdot5 \text{ V}$$

$$= 338 \text{ V}$$

Hence the supply voltage must be of the form $v = 338 \sin(314t - \pi/2)$ V

Note: the current must lead the voltage by $\pi/2$ rad. Since the voltage will be of sine wave form,

$$\text{voltmeter reading} = V = \frac{V_m}{\sqrt{2}} = 0.707 \, V_m$$

$$= 0.707 \times 338 \text{ V} = 239 \text{ V}$$

The voltmeter reading will be 239 V and the instantaneous voltage will be given by $v = 338 \sin(314t - \pi/2)$ V.

Example 9.14

When capacitors of 80 μF and 50 μF are connected in series to a 50 Hz supply the volt drop across the 80 μF capacitor is measured as 98 V. Calculate

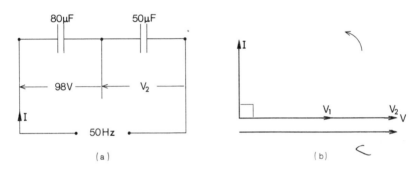

Fig. 9.12 Circuit and phasor diagrams for Example 9.14.

the value of the current in the circuit and the value of the supply voltage. Fig. 9.12(a) and (b) show the circuit and phasor diagrams.

$$\text{Capacitive reactance of 80 μF capacitor} = \frac{10^6}{2\pi f C} = \frac{10^6}{2\pi \times 50 \times 80} \, \Omega$$

$$= 39.8 \, \Omega$$

But   volt drop across capacitance $= IX_C$

Hence

$$I = \frac{98}{39.8} \text{A} = 2.46 \text{ A}$$

$$\text{Capacitive reactance of 50 μF capacitor} = \frac{10^6}{2\pi \times 50 \times 50} \, \Omega = 63.4 \, \Omega$$

Hence volt drop across 50 μF capacitor $= 2.46 \times 63.4$ V $= 156$ V

Since the current must lead on the volt drops across both capacitors by 90°,

143

then the two volt drops must be in phase and so may be added arithmetically.
Hence $\qquad$ supply voltage $= 98 + 156 \text{ V} = 254 \text{ V}$

The current in the circuit will be 2·46 A and the supply voltage is 254 V.
Alternative solutions for determining the supply voltage.

(a) The supply voltage is given by $IX_t = I \times \dfrac{10^6}{2\pi f C_t}$

For capacitors in series, $\qquad C_t = \dfrac{C_1 \times C_2}{C_1 + C_2}$

$$= \dfrac{80 \times 50}{80 + 50}\,\mu\text{F} = \dfrac{400}{13}\,\mu\text{F}$$

Hence $\qquad X_t = \dfrac{10^6 \times 13}{2\pi \times 50 \times 400}\,\Omega = 103\cdot 8\,\Omega$

$$\text{supply voltage} = 2\cdot 46 \times 103\cdot 8 \text{ V} = 255 \text{ V}$$

(b) The supply voltage is given by $\quad V_t = V_1 \times \dfrac{C_1}{C_t} \qquad (V_t \omega C_t = V_1 \omega C_1)$

$$= 98 \times 80 \times \dfrac{13}{400} \text{ V} = 255 \text{ V}$$

Example 9.15
A parallel-plate capacitor is built up of 9 plates, each having an area of 800 mm² and each pair of plates are separated by paper 0·01 mm thick and having a relative permittivity of 2·5. Calculate the current taken by this capacitor when it is connected across a p.d. of 50 mV at 10 kHz.

Since capacitance $\quad C = \dfrac{\epsilon_o \epsilon_r A}{d}$

then $\qquad C = \dfrac{8\cdot 85 \times 10^{-12} \times 2\cdot 5 \times 8 \times 800 \times 10^{-6} \text{ F}}{0\cdot 01 \times 10^{-3}}$

$$= 0\cdot 0142\,\mu\text{F}$$

At a frequency of 10 kHz,

$$X_C = \dfrac{10^6}{2\pi \times 10 \times 10^3 \times 0\cdot 0142}\,\Omega = 1128\,\Omega$$

$$I = \dfrac{V}{X_C} = \dfrac{50 \times 10^{-3}}{1128}\,\text{A}$$

$$= 44\cdot 3\,\mu\text{A}$$

The current taken by the parallel-plate capacitor will be 44·3 μA.

Note: The current *I* taken by a capacitor is given by

$$I = \frac{V}{X_C} = \frac{V}{1/\omega C} = V\omega C.$$

Example 9.16
A 10 μF capacitor takes a current of 1 A from a 250 V sinusoidal supply. What must be the frequency of the supply?

Example 9.17
Determine the equation for the current produced when a 40 μF capacitor is connected to a supply given by $v = 340 \sin 110\pi t$ V

Example 9.18
A capacitor is rated at 60 μF, 110 V, 50 Hz. Calculate the capacitance of the required series capacitor to enable the 60 μF capacitor to operate safely on a 250 V, 50 Hz supply.

Example 9.19
A current, of 6 A, leading on the voltage by $\pi/2$ rad is taken from a 250 V, 50 Hz supply. Calculate the capacitance of the capacitor. If the frequency of the supply were increased to 60 Hz what then would be the current? The supply voltage remains constant.

Example 9.20
A circuit consists of a 120 μF capacitor in series with two 120 μF capacitors connected in parallel. If the total current is 6 A at a frequency of 50 Hz, determine the equation for the supply voltage and the p.d. across each of the capacitors.

9.4 Resistance and Inductance in Series

Fig. 9.13 shows such a circuit. At any instant,

$$v = v_R + v_L$$

Fig. 9.14(a) shows the curves of v_R and v_L drawn with reference to the current

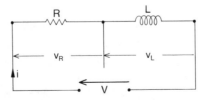

Fig. 9.13 Circuit of resistance and inductance in series.

145

W

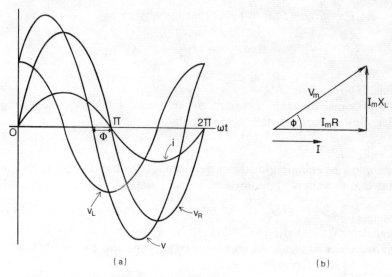

(a)   (b)

Fig. 9.14 (a) Wave forms of current and voltages for circuit in Fig. 9.13.
(b) Phasor diagram for the same circuit.

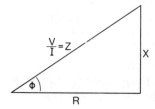

Fig. 9.15 Impedance triangle derived from Fig. 9.14(b).

and then added algebraically to give the voltage curve v. These curves show that the current lags on the voltage by some angle ϕ—called the angle of phase difference—and its value will vary between 0 and $\pi/2$ rad depending on the relative values of R and X_L.

Fig. 9.14(b) shows the addition of phasors to give the applied voltage. The maximum values have been used in this diagram to allow easy comparison to be made with the curves. It is normal practice to use r.m.s. values on phasor diagrams. Since it is a series circuit, the current is the same in all parts of the circuit and therefore may be used as a reference or datum line.

The $I_m R$ phasor is drawn in phase with the current, i.e. it is drawn parallel to the current phasor and to the end of the $I_m R$ phasor is drawn that for $I_m X_L$. Since, in a circuit of pure inductance the current lags by $\pi/2$ rad, then the $I_m X_L$ phasor is drawn as leading (anti clockwise) the $I_m R$ phasor by $\pi/2$ rad.

146

The phasor drawn from the start of the phasor for the resistive drop, to the end of the phasor for the reactive drop, represents the maximum supply voltage V_m, and ϕ is the angle of phase difference.

The triangle, formed by the voltage phasors, is called the voltage triangle and if each side of this triangle is divided by the current—which is constant for the circuit—a similar triangle is produced.

Fig. 9.15 shows this triangle in which the sides are R, X_L, and V/I. The ratio V/I in such a circuit is called the *impedance*; it is measured in ohms and denoted by Z.

$$V/I = Z$$

Since the triangle is right-angled, then

$$Z^2 = R^2 + X_L^2$$

or

$$Z = \sqrt{(R^2 + X_L^2)}$$

and

$$\phi = \cos^{-1}\frac{R}{Z}$$

$$= \sin^{-1}\frac{X_L}{Z} = \tan^{-1}\frac{X_L}{R}$$

Example 9.21
A coil of resistance 5 Ω and inductance 0·1 H is connected in series with a 15 Ω non-inductive resistor to a 240 V, 50 Hz supply. Calculate the total current, its phase angle with respect to the voltage and the p.d. across the coil.

Fig. 9.16(a) and (b) show the circuit and phasor diagrams.

(*Note*: it is always advisable to draw the circuit and phasor diagrams for all a.c. problems and it is often helpful to letter the points on the circuit as this makes reference to those parts so much easier.)

Total resistance of circuit $= (5 + 15)\,\Omega = 20\,\Omega$

Reactance of circuit, $X_L = 2\pi f L = 2\pi \times 50 \times 0{\cdot}1\,\Omega$

$$= 31{\cdot}4\,\Omega$$

$$Z = \sqrt{(R^2 + X_L^2)}$$

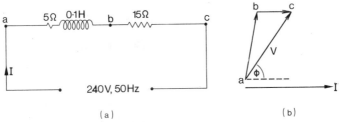

(a)   (b)

Fig. 9.16 Circuit and phasor diagrams for Example 9.21.

147

$$= \sqrt{(20^2 + 31 \cdot 4^2)}\,\Omega$$
$$= \sqrt{(400 + 987)}\,\Omega = 37 \cdot 2\,\Omega$$
$$I = \frac{V}{Z} = \frac{240}{37 \cdot 2}\,\text{A}$$
$$= 6 \cdot 45\,\text{A}$$
$$\cos\phi = \frac{R}{Z} = \frac{20}{37 \cdot 2}$$
$$= 0 \cdot 538$$

Hence
$$\phi = 57 \cdot 5^\circ \text{ (lag)}$$

the p.d. across coil $= {}_aV_b = I_a Z_b$ V

$${}_aZ_b = \sqrt{(5^2 + 31 \cdot 4^2)}/\Omega$$
$$= \sqrt{(25 + 987)}\,\Omega = 31 \cdot 8\,\Omega$$

and so
$${}_aV_b = 6 \cdot 45 \times 31 \cdot 8\,\text{V} = 220\,\text{V}$$

The total current is 6·45 A, lagging on the voltage by 57·5°, and the p.d. across the coil is 220 V.

Example 9.22
When a circuit is connected to an alternating supply given by $v = 340 \sin 314t$ V the resulting current is $i = 17 \sin(314t - \pi/8)$ A. Determine the impedance, the resistance and the inductance of the circuit. Estimate the value of the current 1·5 ms after the instant that the supply is zero and increasing positively.

$$\text{Impedance } Z = \frac{V}{I} = \frac{V_m}{I_m}$$
$$= \frac{340}{17}\,\Omega = 20\,\Omega$$

Since
$$\cos\phi = R/Z$$
then
$$R = Z\cos\phi$$

From the current equation the current is lagging by $\pi/8$ rad and so

$$\phi = \frac{\pi}{8}\,\text{rad} = 22 \cdot 5^\circ$$

Hence
$$R = 20\cos 22 \cdot 5^\circ\,\Omega$$
$$= 20 \times 0 \cdot 924\,\Omega = 18 \cdot 5\,\Omega$$
$$L = X_L/\omega$$
but
$$\sin\phi = X_L/Z$$
and so
$$X_L = 20\sin 22 \cdot 5^\circ\,\Omega$$
$$= 20 \times 0 \cdot 383\,\Omega = 7 \cdot 66\,\Omega$$

$$L = \frac{7 \cdot 66}{314} H = 24 \cdot 4 \, mH$$

When $t = 1 \cdot 5$ ms,

$$i = 17 \sin \left(314 \times 1 \cdot 5 \times 10^{-3} - \frac{\pi}{8} \right) A$$
$$= 17 \sin (0 \cdot 471 - 0 \cdot 393) \, A$$
$$= 17 \sin 0 \cdot 078 \, A$$
$$= 17 \times 0 \cdot 0785 \, A = 1 \cdot 335 \, A$$

Instead of using the angles in radians they could be converted to degrees.

$$i = 17 \sin (27° - 22 \cdot 5°) \, A$$
$$= 17 \sin 4 \cdot 5° \, A$$
$$= 17 \times 0 \cdot 0785 \, A$$
$$= 1 \cdot 335 \, A$$

The impedance of the circuit is 20 Ω, the resistance 18·5 Ω and the inductance 24·4 mH, and after 1·5 ms the current will be 1·34 A.

Example 9.23
A 240 V, 50 Hz supply is connected to a circuit consisting of a 10 Ω resistor in series with a reactor, having a resistance of 3 Ω. If a current of 10 A flows in the circuit determine, by means of a phasor diagram, the volt drop across the reactor and calculate its inductance.

Fig. 9.17(a) and (b) show the circuit diagram and a sketch of the phasor

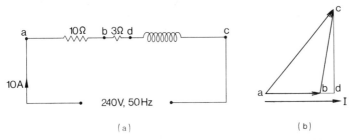

Fig. 9.17 Circuit and phasor diagrams for Example 9.23.

diagram. (It is always helpful to sketch the diagram before drawing it to scale.)

$$\text{p.d. across resistor} = IR$$

and so
$$_aV_b = 10 \times 10 \, V = 100 \, V$$

This will be in phase with the current.

149

The volt drop across the reactor will have two components: an in-phase component (Ir) and a quadrature component (IX_L)

$$\text{In-phase component} = 10 \times 3 \text{ V} = 30 \text{ V}$$

Choose a suitable scale for the phasor diagram, e.g. 1 mm ≡ 2 V. Fig 9.18 shows the phasor diagram drawn to this scale. For a series circuit the current is the reference. Draw a line ab parallel to the current phasor and of length 50 mm, and at b add the in-phase component of the volt drop across the reactor, of length 15 mm. Let this be bd.

At d draw a perpendicular (anticlockwise) and from a draw an arc of radius 120 mm to cut the perpendicular at c. Then bc ≡ the volt drop across the reactor.

$$\text{Length of } bc = 101 \text{ mm}$$

Hence
$$\text{p.d. across reactor} = 101 \times 2 \text{ V} = 202 \text{ V}$$

$$\text{Length of } dc = 100 \text{ mm}$$

Hence
$$IX_L = 100 \times 2 \text{ V}$$

and so
$$I\omega L = 200 \text{ V}$$

$$L = \frac{200}{2\pi \times 50 \times 10} \text{ H} = 64 \text{ mH}$$

By measurement the p.d. across the reactor is 202 V and its inductance is 64 mH.

Example 9.24
A 50 Hz, sinusoidal current of 4 A flows in a circuit of resistance 24 Ω and inductive reactance 18 Ω. For a complete cycle draw the curves of the resistive and reactive volt drops. From these draw the curve for the applied voltage and so obtain the equation for the applied voltage.

Example 9.25
A coil, having a resistance of 5 Ω and an inductance of 38·2 mH, is connected across a 50 Hz supply of 130 V. Calculate the current taken from the supply and its phase angle.

Example 9.26
A non-inductive resistance of 15 Ω is connected in series with a coil of resistance 9 Ω and inductance 57·3 mH to a 240 V, 50 Hz supply. Determine the r.m.s. value of the current, its phase angle and the p.d. across the coil.

Example 9.27
Two coils A and B are connected in series to a 250 V, 50 Hz supply. Coil A has a resistance of 5 Ω and an inductance of 116 mH and coil B has a resistance

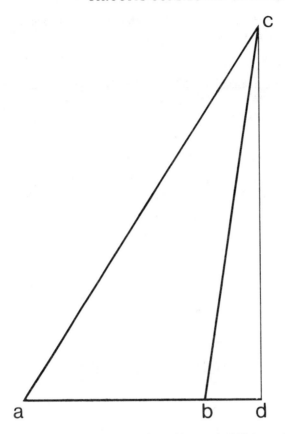

Fig. 9.18 Phasor diagram for solution of Example 9.23 (to scale).

of 40 Ω and an inductance of 75 mH. Determine the r.m.s. value of the current, its phase angle and the p.d. across each coil.

Example 9.28
When a coil is connected in series with a 20 Ω resistor to a 240 V, 50 Hz supply a current of 6 A is drawn from the supply. Estimate the inductance of the coil if its resistance is 4 Ω.

Example 9.29
A coil takes a current of 10 A at a lagging phase angle of 37° from a 240 V, 50 Hz supply. Determine the resistance and the inductance of the coil.

Example 9.30
A coil takes 8 A from a 240 V, 50 Hz supply. When the frequency of the supply

is increased to 100 Hz the current is reduced to 6 A. Calculate the resistance and inductance of the coil.

Example 9.31
A coil takes a current of 5 A from a 30 V d.c. supply. When the same coil is connected to a 240 V, 50 Hz supply the current is only 3·6 A. Determine the inductance of the coil.

9.5 Resistance and Capacitance in Series

Fig. 9.19 shows such a circuit in which a current $i = I_m \sin \omega t$ is flowing. In a series circuit the current is the same in all parts of the circuit and the

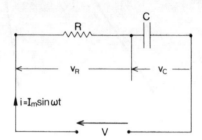

Fig. 9.19 Circuit of resistance and capacitance in series.

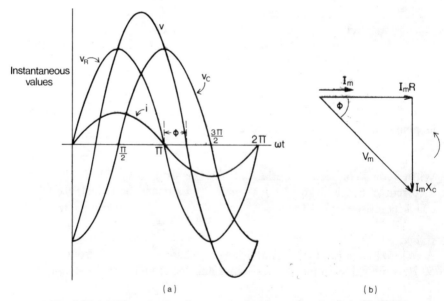

Fig. 9.20 (a) Wave forms of current and voltages for circuit in Fig. 9.19.
(b) Phasor diagram for the same circuit.

152

applied voltage will be equal to the phasor sum of the separate voltage drops.

Fig. 9.20(a) shows the curves of these voltage drops. The resistive volt drop v_R will be in phase with the current and the reactive volt drop v_C will be in quadrature with the current, the current leading by $\pi/2$ rad or 90°. These curves have been drawn for the special case when $R = X_C$. This allows easier comparison with the phasor diagram shown in Fig. 9.20(b)—drawn to maximum values. By adding the curves of v_R and v_C the curve for the applied voltage may be found and the curves in Fig. 9.20(a) show that the current leads on the voltage by some angle ϕ. In this special case $\phi = \pi/4$ rad. Comparison with the phasor diagram will show that the latter gives as much information as do the curves.

From the voltage phasor triangle in Fig. 9.20(b),

$$V_m^2 = (I_m R)^2 + (I_m X_C)^2$$

Converting to r.m.s. values,

$$V = \sqrt{\{(IR)^2 + (IX_C)^2\}}$$
$$= I\sqrt{\{R^2 + X_C^2\}}$$

Hence

$$\frac{V}{I} = Z = \sqrt{\{R^2 + X_C^2\}}$$

and as with resistance and inductance in series,

$$\cos\phi = R/Z$$

Example 9.32

A 60 Ω resistor is connected in series with a 40 µF capacitor to a 250 V, 50 Hz supply. Calculate the current taken from the supply and the angle of phase difference.

Fig. 9.21 shows the circuit and phasor diagrams.

$$\text{Capacitive reactance } X_C = \frac{1}{\omega C} = \frac{10^6}{2\pi \times 50 \times 40}\,\Omega$$

$$= 80\,\Omega$$

$$\text{Impedance of circuit, } Z = \sqrt{(R^2 + X_C^2)}$$

$$= \sqrt{(60^2 + 80^2)}\,\Omega = 100\,\Omega$$

But

$$I = \frac{V}{Z} = \frac{250}{100}\,\text{A}$$

$$= 2 \cdot 5\,\text{A}$$

$$\text{Phase angle } \phi = \cos^{-1}\frac{R}{Z} = \cos^{-1}\frac{60}{100}$$

$$= \cos^{-1} 0 \cdot 6 = 53°\,\text{(lead)}$$

153

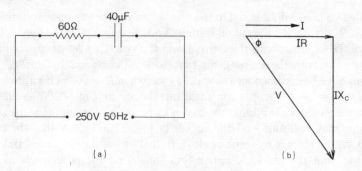

Fig. 9.21 Circuit and phasor diagrams for Example 9.32.

The current taken from the supply will be 2·5 A and it will lead on the voltage by 53°.

Example 9.33
A 25 Ω resistor is connected in series with a 100 μF capacitor to a 240 V, 50 Hz supply. Calculate the current and its phase angle.

Example 9.34
A 20 Ω resistor is rated to carry 5 A. Determine the value of capacitance to be connected in series with the resistor so that it may be connected to a 240 V, 50 Hz supply.

Example 9.35
Two capacitors are to be connected in series to a 250 V, 50 Hz supply. One capacitor has a capacitance of 100 μF. Calculate the capacitance of the second capacitor so that a current of 5 A shall be taken from the supply.

Example 9.36
The instantaneous value of voltage is represented by $v = 360 \sin 314t$ V and that of the current by $i = 45 \sin(314t + 2\pi/9)$ A. Calculate the resistance, the reactance and the impedance of the circuit. What will be the value of the current at the instant that the voltage is 200 V and increasing positively?

Example 9.37
A current of 5 A, having a sine wave form and a frequency of 50 Hz, flows in a circuit of resistance 3 Ω and capacitive reactance 4 Ω. Draw curves of the resistive and the reactive voltage drops and from these obtain the curve for the applied voltage. What is the maximum value of this applied voltage and what is the phase angle of the circuit?

154

9.6 Power in a.c. Circuits

(a) Pure resistance. At any instant the power in a circuit is the product of the current and voltage at that instant.

$$p = iv$$

Fig. 9.22(a) shows the curves for the current and voltage in a circuit of pure resistance and if these curves are multiplied together a wholly positive curve will result. Fig. 9.22(b) shows the resultant curve; it is a sine wave of double the original frequency and has an average value of $V_m I_m / 2$. Hence

$$\text{average power} = \frac{V_m I_m}{2} = \frac{V_m}{\sqrt{2}} \times \frac{I_m}{\sqrt{2}}$$

$$= V \times I$$

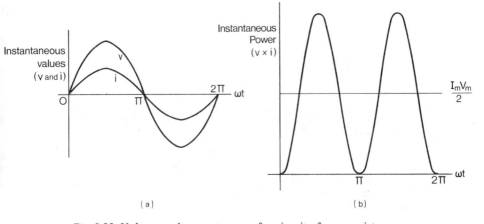

(a)   (b)

Fig. 9.22 Voltage and current curves for circuit of pure resistance.

In a circuit of pure resistance, in which wave forms are sinusoidal, the average power is equal to the product of the r.m.s. voltmeter reading and the r.m.s. ammeter reading. This is the power indicated on a wattmeter.

(b) Pure reactance. In a circuit of pure reactance (inductive or capacitive) the current and the voltage are in quadrature, and when the instantaneous values of the current and voltage are multiplied together, the sine wave produced is of double the original frequency and is symmetrical about the zero axis.

Fig. 9.23 shows this power curve. Since it is symmetrical about the zero axis then the average power in a circuit of pure reactance is zero.

(c) Resistance and reactance in series. Fig. 9.24(a) and (b) show the voltage phasor triangles for circuits of resistance in series with (a) inductive reactance

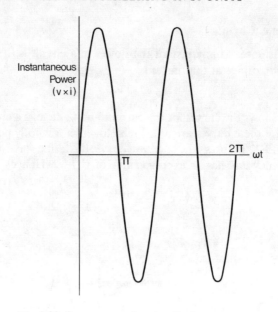

Fig. 9.23 Power curve for circuit of pure reactance.

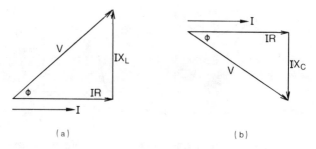

Fig. 9.24 Voltage triangles for circuits of resistance and reactance in series.

and (b) capacitive reactance. If each side of these triangles is multiplied by the current, power triangles will be produced.

IR is the volt drop across the resistive part of the circuit, and so

$$IR \times I = I^2R$$

This is the true power in the circuit. But

$$IR = V \cos \phi$$

and so

$$P = VI \cos \phi$$

and is measured in watts (W) or in kilowatts (kW)

Also,

$$IX \times I = V \sin \phi \times I = VI \sin \phi$$

156

This is called the reactive power and is measured in volt-amperes reactive (var) or kilovolt-amperes reactive (kvar).

The product of the supply voltage and the current is called the apparent power and is measured in volt amperes or kilovolt amperes (kVA).

The ratio between the true power and the apparent power is called the power factor.

$$\text{Power factor} = \frac{\text{true power}}{\text{apparent power}}$$

$$= \frac{VI \cos \phi}{VI}$$

$$= \cos \phi \quad \text{(for sine waveforms only)}$$

Example 9.38
The element of an electric radiator, rated at 240 V, 1 kW, is non-inductively wound. Calculate the current the element will take when operating correctly.

Since the element is non-inductively wound then its impedance is purely resistive. Hence

$$P = VI$$

$$I = \frac{P}{V} = \frac{1000}{240} \text{ A}$$

$$= 4 \cdot 17 \text{ A}$$

The current taken by the element will be 4·17 A.

Example 9.39
A current of 5 A flows in a circuit having a resistance of 8 Ω and an inductance of 80 mH. Determine the power converted into heat in the circuit.

Since power is only consumed in the resistance of the circuit, then

$$P = I^2 R = 5^2 \times 8 \text{ W}$$

$$= 200 \text{ W}$$

The amount of power converted into heat is 200 W.

Example 9.40
Calculate the impedance, resistance, reactance and power factor of an iron-cored coil which takes a current of 12 A and dissipates 600 W when it is connected to a 240 V, 50 Hz sine-wave supply.

$$\text{Impedance of coil, } Z = \frac{V}{I} = \frac{240}{12} \Omega$$

$$= 20 \Omega$$

Since $$R = P/I^2 \quad \text{and} \quad P = 600 \text{ W}, \quad \text{then}$$

$$R = \frac{600}{12^2}\,\Omega = 4\cdot17\,\Omega$$

Hence

$$\text{power factor, } \cos\phi = \frac{R}{Z} = \frac{4\cdot17}{20}$$

$$= 0\cdot209 \text{ (lagging)}$$

reactance, $$X = \sqrt{(Z^2 - R^2)}\,\Omega$$

$$= \sqrt{(20^2 - 4\cdot17^2)}\,\Omega = 19\cdot56\,\Omega$$

The impedance, resistance and reactance of the coil are 20 Ω, 4·17 Ω and 19·56 Ω respectively and its power factor is 0·209 (lagging).

Note: When the power dissipated by a coil is divided by the current squared, the value of resistance so obtained is called the effective, or a.c., resistance of the coil. This value of resistance is generally greater than the d.c. resistance due to the fact that power losses may occur in the iron core.

Example 9.41
When the output of a 230 V, 50 Hz, a.c. motor is 3·8 kW, its per-unit efficiency and power factor are 0·85 and 0·88 respectively. Determine the true power, apparent power and reactive power taken from the supply.

$$\text{Output of motor} = 3\cdot8 \text{ kW}$$

Since $$\text{input} = \text{output/efficiency}$$

then $$\text{power input} = 3\cdot8/0\cdot85 \text{ kW}$$

$$= 4\cdot47 \text{ kW} \quad \text{(this is the true power)}$$

But $$\text{power factor} = \text{true power/apparent power}$$

and so $$\text{apparent power} = \text{true power/power factor}$$

$$= \frac{4\cdot47}{0\cdot88} \text{ kVA} = 5\cdot07 \text{ kVA}$$

$$\text{Reactive power} = VI\sin\phi$$

$$= \text{apparent power} \times \sin\phi$$

Here $$\cos\phi = 0\cdot88$$

$$\phi = 28°\,20'$$

and so $$\sin\phi = 0\cdot4746$$

Hence $$\text{reactive power} = 5\cdot07 \times 0\cdot4746 \text{ kv ar}$$

$$= 2\cdot4 \text{ kv ar}$$

Alternative solution for reactive power. From the power triangle,

$$VA = \sqrt{\{(W)^2 + (var)^2\}}$$

and so

$$var = \sqrt{\{(VA)^2 - (W)^2\}}$$
$$kvar = \sqrt{(5 \cdot 07^2 - 4 \cdot 47^2)}$$
$$= \sqrt{(25 \cdot 7 - 19 \cdot 98)}$$
$$= \sqrt{5 \cdot 72} = 2 \cdot 392$$

The true power taken from the supply is 4·47 kW, the apparent power is 5·07 kVA and the reactive power is 2·4 kvar.

Example 9.42
A 15 Ω resistor is connected to a 240 V, 50 Hz supply. Calculate the power dissipated in the resistor.

Example 9.43
When a voltage of the form $v = 340 \sin 314t$ V is connected to an impedance the resulting current is of the form $i = 17 \sin 314t$ A. Determine the power dissipated in the impedance.

Example 9.44
A coil, having a lagging power factor of 0·6, takes a current of 16 A from a 240 V, 50 Hz supply. Calculate the effective resistance and inductance of the coil.

Example 9.45
A voltage, of the form $v = 354 \sin 500t$ V, is applied to an impedance and the resulting current is of the form $i = 17 \cdot 7 \sin(500t - 37°)$ A. Determine the true power and the apparent power taken from the supply. Is the impedance inductive or capacitive? Calculate the inductance or the capacitance of the impedance.

Example 9.46
Estimate the output of an a.c. motor which takes 11 kW at a lagging power factor of 0·9 from a 240 V, 50 Hz supply if the efficiency of the motor is 0·8. Calculate the load current, the apparent power and the reactive power.

Example 9.47
A purely resistive appliance operates correctly when taking 2·5 A from a 110 V supply. It can be made to operate correctly on a 240 V, 50 Hz supply by connecting it in series with either a resistor or a reactor. Calculate the value of series resistance required and the power dissipated in it. How much power

would be saved by using a reactor having a power factor of 0·4? (Draw, to scale, the phasor diagram for the second case and so determine the p.d. across the reactor.)

Chapter 10

Electronics

10.1 The Thermionic Diode

This is so called because it is a two electrode valve and electron emission is obtained by direct or indirect heating of the cathode. Its main uses are (a) as a rectifier, (b) as a detector.

Example 10.1
In a test to determine the static anode characteristic the following results were obtained:

Anode voltage (V_a) in V	0	25	37·5	50	62·5	75	125	150	175	200	250
Anode current (I_a) in A	0	0·75	1·25	2·5	3·5	5·0	11·3	13·8	15	16	16·2

From this characteristic obtain the anode a.c. resistance for an anode potential of 150 V and the anode d.c. resistance for an anode potential of 120 V.

Fig. 10.1 shows the characteristic drawn to the following scales:

$$\text{Horizontal scale: Anode potential } 1 \text{ mm} \equiv 0\cdot5 \text{ V}$$

$$\text{Vertical scale: Anode current} \quad 1 \text{ mm} \equiv 0\cdot25 \text{ mA}$$

At the point on the characteristic corresponding to an anode potential of 150 V draw a straight line tangential to the characteristic. The slope of this tangent is the a.c. conductance of the diode. From Fig. 10.1 the slope of the tangent is

$$\frac{17\cdot5 - 6\cdot25}{200 - 50} \frac{\text{mA}}{\text{V}} = \frac{11\cdot25}{150} \text{ mS}$$

Hence $\quad$ a.c. resistance of diode $= \dfrac{150}{11\cdot25} \text{ k}\Omega = 13\cdot3 \text{ k}\Omega$

To determine the d.c. anode resistance, join the point on the characteristic,

x

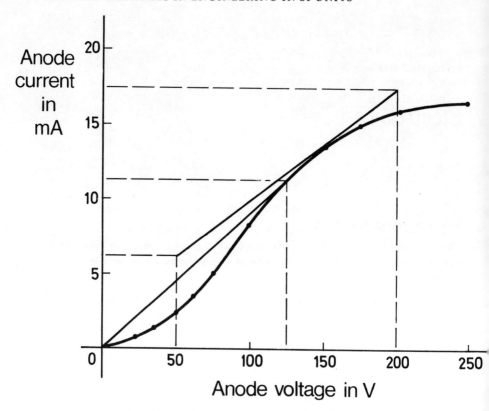

Fig. 10.1 Anode characteristic for Example 10.1 (to scale).

corresponding to the anode potential of 120 V, to the origin. The slope of this line is equal to the reciprocal of the d.c. anode resistance. From Fig. 10.1,

$$\text{slope of line to origin} = \frac{11{\cdot}25 - 0}{120 - 0} \frac{\text{mA}}{\text{V}}$$

Hence   d.c. anode resistance of diode $= \dfrac{120}{11{\cdot}25} \, \text{k}\Omega = 10{\cdot}7 \, \text{k}\Omega$

For the stated potentials the a.c. and d.c. anode resistances are 13·3 kΩ and 10·7 kΩ respectively.

Example 10.2
A sinusoidal voltage, having a peak value of 350 V and a period of 20 ms, is applied to a circuit consisting of a diode in series with a resistor across which is connected a reservoir capacitor. Determine the mean d.c. voltage developed

162

across the resistor. The *RC* circuit has a time constant of 100 ms and linear discharge of the capacitor may be assumed. Assume the diode has negligible forward resistance and infinite reverse resistance.

Fig. 10.2(a) shows the circuit and Fig. 10.2(b) shows the rectified wave form which would be developed across the resistor if no capacitor were connected. Scales for Fig. 10.2(b) are as follows:

Horizontal scale: Time $\qquad$ 6 mm $\equiv$ 1 ms

Vertical scale: Instantaneous voltages $\quad$ 1 mm $\equiv$ 5 V

With the capacitor in circuit, during the first positive half-cycle the capacitor will become charged to the peak a.c. voltage, and as soon as the p.d. across the resistor falls below that across the capacitor, the capacitor will discharge through the resistor.

In a time interval equal to the time constant the capacitor p.d. will decrease by $0{\cdot}632 \times 350$ V, i.e. by 221 V. Hence the linear rate of voltage decrease will be $221/100 = 2{\cdot}21$ V/ms, so that in 20 ms the capacitor p.d. will decrease by $2{\cdot}21 \times 20$ V = $44{\cdot}2$ V.

At a time of 5 ms the capacitor p.d. is 350 V, and so at a time of 25 ms the capacitor p.d. would be $350 - 44{\cdot}2$ V = $305{\cdot}8$ V. Fig. 10.2(c) shows this point marked on the diagram and a straight line is drawn from this point to the apex of the first positive half-cycle. This line represents the variation of p.d. across the resistor and it will continue until the line meets the rising p.d. of the next positive half-cycle, at 310 V. Hence the p.d. across the resistor will consist of a mean d.c. voltage on which there is superimposed a saw tooth of alternating wave form. From Fig. 10.2(c),

$$\text{peak voltage of a.c. wave} = \frac{350 - 310}{2} \text{ V} = 20 \text{ V}$$

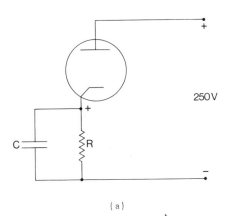

(a)

Fig. 10.2 (a) Circuit diagram for Example 10.2.
$\qquad$ (b) Voltage curve across resistor without capacitor in circuit (to scale).
$\qquad$ (c) Curve of voltage across RC circuit (to scale).

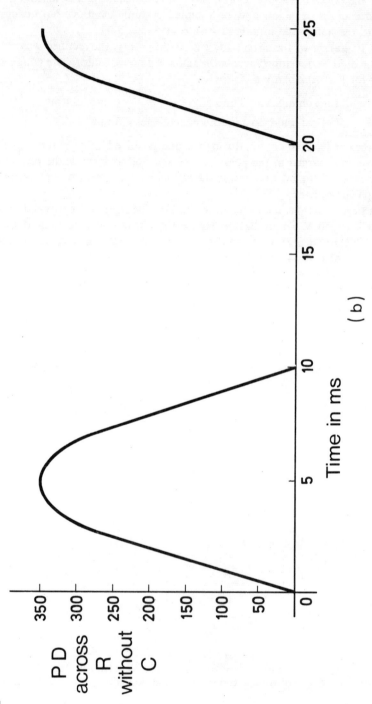

(b)

Time in ms

PD across R without C

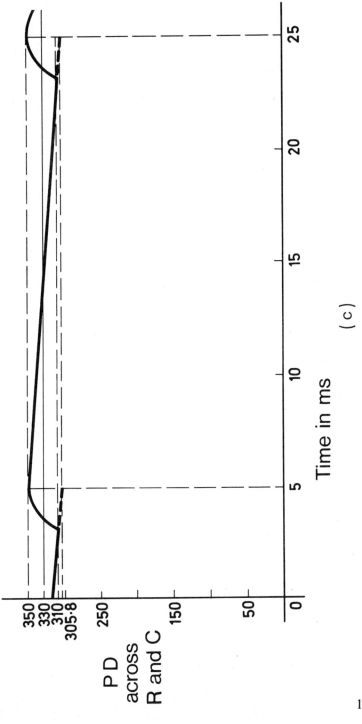

Time in ms

(c)

PD across R and C

350
330
310
305·8
250
150
50
0

and so mean voltage across the resistor $= 350 - 20$ V $= 330$ V

The mean voltage across the resistor is 330 V.

Example 10.3
A load test performed on a diode with reservoir capacitor gave the following results:

A.C. input voltages in V	Load currents in mA	D.C. output								
		0	5	10	15	20	30	40	50	60
250	Load p.d. in V	345	325	307	292	280	258	240	220	202
220	Load p.d. in V	310	290	270	255	245	220	202	188	172
200	Load p.d. in V	280	258	240	225	215	190	177	165	145

From these characteristics estimate for an a.c. input of 250 V, the load current for a d.c. voltage of 250 V, and estimate the required a.c. input voltage to give a load current of 45 mA at a d.c. voltage of 220 V.
Fig. 10.3 shows the characteristics drawn to the following scales:

Horizontal scale: Load currents $\quad$ 1 mm $\equiv$ 0·5 mA

Vertical scale: D.C. output voltage $\quad$ 1 mm $\equiv$ 2·5 V

Draw a horizontal line through the 250 V d.c. to cut the 250 V a.c. characteristic. Where the line cuts the characteristic represents the load current at 250 V d.c.

From Fig. 10.3 the load current at 250 V d.c. is 34·5 mA. Draw a vertical line through the point corresponding to a load current of 45 mA to cut all the characteristics. Read off from the curves the d.c. voltages at which the characteristics are cut, i.e. 172, 196 and 230 V. Plot these values against the corresponding a.c. input voltages.
Fig. 10.4 shows this curve plotted to the following scales:

Horizontal scale: D.C. output voltages $\quad$ 1 mm $\equiv$ 1 V

Vertical scale: A.C. input voltages $\quad$ 1 mm $\equiv$ 1 V

The required a.c. voltage can then be read off the curve for a d.c. output voltage of 220 V. From the curve the required a.c. voltage is 238 V.
Note: An approximate value may be obtained by proportion. The vertical intercept between the 250 V and the 220 V characteristics is 14 mm.
For 45 mA at 220 V d.c. the point is 9 mm above the 220 V characteristic, and so

$$\text{required a.c. voltage} = 220 + 30 \times \frac{9}{14} \text{ V}$$

$$= 220 + 19\cdot3 \text{ V} = 239 \text{ V}$$

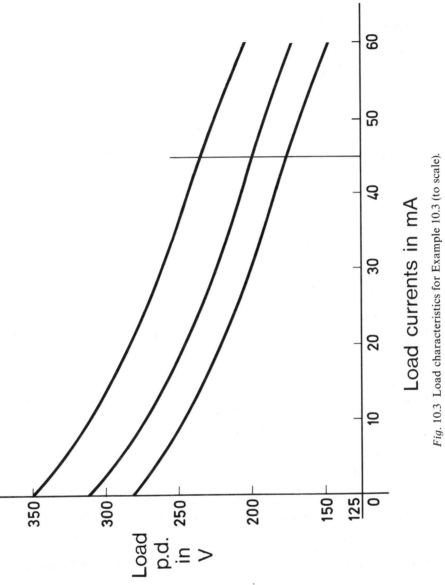

Fig. 10.3 Load characteristics for Example 10.3 (to scale).

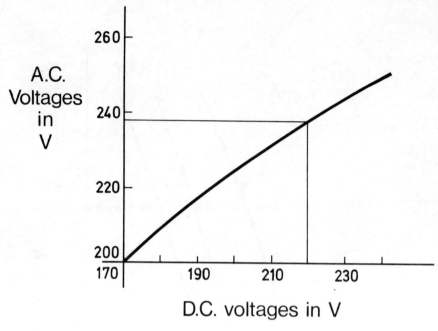

Fig. 10.4 Curve of a.c. input/d.c. output for Example 10.3 (to scale).

The maximum load current at 250 V d.c. is 35 mA and the estimated a.c. input voltage is 239 V.

Example 10.4
When a diode is connected in series with a resistor to a 100 V d.c. supply the current in the circuit is 5·35 mA. If the characteristic of the diode is as given below, determine the value of the series resistance.

Anode voltage V_a in V	0	10	20	30	40	60	80
Anode current I_a in mA	0	0·2	0·95	2·2	4·0	8·0	9·8

Fig. 10.5 shows the diode characteristic drawn to the following scales:

Horizontal scale: Anode voltage V_a 1 mm ≡ 1 V
Vertical scale: Anode current I_a 1 mm ≡ 0·1 mA

Since the current in the circuit is 5·35 mA, the operating point on the characteristic is at *L*.

The value of the series resistance may be determined either (a) by calculation or (b) graphically.

(a) From the characteristic at an anode current of 5·35 mA the anode voltage is 46 V. Hence

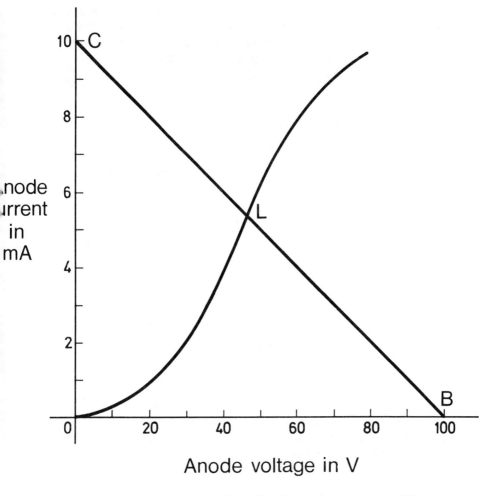

Fig. 10.5 Anode characteristic and load line for Example 10.4 (to scale).

$$\text{volt drop across resistor} = I_a R = 100 - 46 \text{ V}$$
$$= 54 \text{ V}$$

Hence
$$R = \frac{54}{5 \cdot 35 \times 10^{-3}} \Omega = 10 \cdot 1 \text{ k}\Omega$$

(b) If the anode voltage were 100 V then the volt drop across the resistor would be zero and the anode current would be zero (limiting case). This condition is represented on Fig. 10.5 by the point *B*. The point *L* represents the working condition of the circuit. If a straight line is drawn through *B* and *L* then this will represent the variation of the volt drop across the resistor between these limits. If this line is continued to cut the vertical axis at the point C, then point C will represent the condition when the 100 V is being dropped across the resistor. The value of resistance may then be determined. From Fig. 10.5 the point C corresponds to a current of 10 mA.

$$R = \frac{100}{10 \times 10^{-3}} \Omega = 10 \, k\Omega$$

The value of the series resistance is 10 kΩ.

Example 10.5
A thermionic diode has a static anode characteristic given by the following values:

Anode voltage V_a in V	0	15	25	50	100	150	175	200	250	300
Anode current I_a in mA	0	1·5	3·2	8·0	16·5	25·0	28·0	30·0	31·0	31·5

Determine the a.c. resistance of the valve at an anode potential of 125 V and the d.c. resistance at an anode potential of 160 V. What anode voltage is required to produce an anode current of 20 mA?

Example 10.6
A thermionic diode produces a saturation current of 25 mA for an anode voltage of 250 V. What then is the rate at which the cathode is emitting electrons?

Example 10.7
Other factors remaining constant, the saturation current of a thermionic valve is directly proportional to the square of the heater current. If in a valve a heater current of 0·3 A produces a saturation current of 20 mA, what will be the saturation current for a heater current of 0·25 A? What heater current would be required for the anode current to saturate at 25 mA?

Example 10.8
A thermionic diode, with reservoir capacitor, gave the following values for a load test:

A.C. input voltages in V		D.C. Output					
	Load currents in mA	0	12·5	25	50	75	100
240	Voltages in V	330	305	285	265	245	228
220	Voltages in V	305	280	265	242	222	203

For an a.c. input voltage of 240 V, estimate the value of the load resistance so that the load current will be 60 mA. If the input voltage decreased to 220 V to what value would the load resistance have to be altered so that the p.d. across it is 225 V?

170

Example 10.9
A diode valve is connected in series with a 7·5 kΩ resistor to a 250 V d.c. supply. If the diode has the following anode characteristic, estimate the current in the circuit and the p.d. across the valve.

Anode voltage V_a in V	0	12·5	25	37·5	50	75	100	125	175
Anode current I_a in mA	0	1·0	3·5	7·5	12	20	27·5	32	35

Example 10.10
A diode with smoothing capacitor is connected to a supply of the form $v = 150 \sin 314t$ V. If linear discharge of the capacitor may be assumed, estimate graphically the value of the circuit time constant so that the mean d.c. voltage across the resistor is 140 V. Assume the diode has negligible forward resistance and infinite reverse resistance.

10.2 Semiconductor Diodes

A semiconductor is a material with a conductivity that lies between that of a conductor and that of an insulator.

Pure silicon and germanium are both insulators and to convert them to semiconductors it is necessary to add impurities (doping—1 part in 10^6). Both silicon and germanium have valencies of 4 and the addition of elements having a valency of 5, e.g. phosphorus, produces a surplus of electrons—'n-type', but the addition of elements having valencies of 3, e.g. arsenic, produces a shortage of electrons—'p-type'.

A semiconductor diode is produced by combining n-type with p-type to produce a p–n junction. This type of diode has the following advantages over the thermionic diode:

(a) It is smaller and more compact.
(b) It operates without a heated cathode so that less power is used.
(c) Being solid, it is more shock resistance.
(d) It has a higher efficiency.
(e) The cost of production is less.

However, it has the disadvantages that it is damaged by high temperatures and by high reverse voltages.

Example 10.11
In a test on a germanium diode the following forward and reverse characteristics were obtained:

Voltage in V	0	0·22	0·33	0·37	0·41	0·48	0·53	0·59	0·70	0·98	1·13	1·24
Current in mA	0	0·25	1·0	1·5	2·0	3·0	4·0	5·0	7·0	15	20	25

171

Voltage in V	0	2·0	6·0	15·0	24·0	32·0	38·0	43·0	47·0	50	54·5
Current in mA	0	0·03	0·08	0·18	0·3	0·4	0·5	0·6	0·7	0·8	1·0

Plot the characteristics and determine the values of the forward and reverse a.c. resistances at $+1·0$ V and -40 V respectively.

Fig. 10.6 shows the characteristics, different scales being used for the forward and reverse characteristics.

Scales for the forward characteristic are:

$$\text{Horizontal scale: Voltage 1 mm} \equiv 0·02 \text{ V}$$

$$\text{Vertical scale: Current} \quad 1 \text{ mm} \equiv 0·5 \text{ mA}$$

Scales for the reverse characteristic are:

$$\text{Horizontal scale: Voltage 1 mm} \equiv 1 \text{ V}$$

$$\text{Vertical scale: Current} \quad 1 \text{ mm} \equiv 0·02 \text{ mA}$$

To determine the forward and reverse a.c. resistances of the diode draw tangents to the characteristics at the points indicated. From Fig. 10.6,

$$\text{forward resistance} = \frac{1·38 - 0·72}{25 - 5} \frac{\text{V}}{\text{mA}}$$

$$= \frac{0·66}{20 \times 10^{-3}} \Omega = 33\Omega$$

$$\text{reverse resistance} = \frac{60 - 14}{0·96 - 0·0} \frac{\text{V}}{\text{mA}}$$

$$= \frac{46}{0·96 \times 10^{-3}} \Omega = 48·0 \text{ k}\Omega$$

Compare the forward a.c. resistance of the semiconductor diode with the anode a.c. resistance of the thermionic diode in Example 10.1.

Example 10.12

In a laboratory test on a silicon diode the following readings were obtained:

Forward characteristic

P.d. across diode in V	0	0·17	0·19	0·21	0·24	0·28	0·36
Current in mA	0	0·1	0·2	0·3	0·5	1·0	3·0

P.d. across diode in V	0·42	0·52	0·64	0·74	0·81	0·88
Current in mA	5·0	10·0	20·0	30·0	40·0	50·0

Reverse characteristic

P.d. across diode in V	0	14	27	50	59	65·5	69·5	71
Current in μA	0	5	10	30	50	70	90	100

Draw the characteristics and estimate the power dissipated in the diode by a current of 35 mA (I^2r). What is the reverse slope resistance at a p.d. of 60 V?

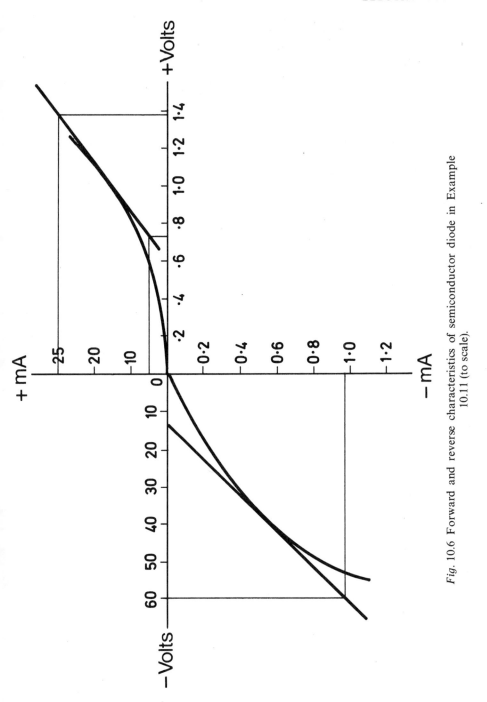

Fig. 10.6 Forward and reverse characteristics of semiconductor diode in Example 10.11 (to scale).

Compare the forward and reverse a.c. resistances with those for the germanium diode in Example 10.11.

Example 10.13
An alternating voltage, having an r.m.s. value of 9 V and a period of 2 ms, is applied to a circuit consisting of a semiconductor diode in series with a resistor across which is connected a 'smoothing' capacitor. If the time constant of the RC circuit is 10 ms, estimate the mean d.c. voltage across the resistor. It may be assumed that the discharge of the capacitor, over the initial 10 ms is linear. To what value would this mean d.c. voltage rise if a second diode were used to rectify the second half cycle?

Example 10.14
A perfect diode in series with a resistor R is connected to a supply of the form $v = 14 \sin 314t$ V. Estimate the mean and r.m.s. values of the voltage across the resistor R.

Example 10.15
In a diode valve voltmeter the scale reading is proportional to the square of the current in the instrument. If a current of 4 μA gives a scale reading of 60 V, what current is needed to give a scale reading of 80 V? What is the scale reading when the current is 6 μA?

Chapter 11

Some Electrical Measuring Instruments

11.1 The Permanent-magnet Moving-coil Instrument

Torque is produced by the interaction of the flux produced by the coil, when carrying a current, and the flux of the 'aged' permanent magnet. 'Ageing' of a magnet is the artificial removal of the temporary magnetism to ensure a constant flux during the working life of the instrument.

The control torque is provided by two spiral springs. The two springs are wound in opposite directions to counteract the effect of expansion due to an increase in temperature.

Example 11.1
The coil of a permanent-magnet moving-coil (p.m.m.c.) instrument consists of 20 turns wound on an aluminium former measuring 30 mm × 20 mm, the longer side being along the axis of rotation. The control springs exert a torque of 50 μNm for full scale deflection (f.s.d.). If the flux density produced by the permanent magnet is 0·12 T, determine the current required to give f.s.d.

The force on a current carrying conductor is given by $F = BIl$. Hence

$$\text{force on one conductor} = 0\text{·}12I \times 0\text{·}03 \text{ N}$$
$$\text{force per coil side} = 20 \times 3\text{·}6 \times 10^{-3}I \text{ N}$$
$$= 72 \times 10^{-3}I \text{ N}$$
$$\text{torque per coil side} = 72 \times 10^{-3} \times \frac{0\text{·}02}{2} I \text{ Nm}$$
$$= 720I \text{ μNm}$$
$$\text{total torque} = 2 \times 720I \text{ μNm}$$

But   total torque for f.s.d. $= 50 \text{ μNm}$

and so $1440I = 50$

$$I = \frac{50}{1440} \text{ A}$$

$$= 34\cdot7\,\text{mA}$$

The current required for f.s.d. is 34·7 mA.

Note: Since

$$F = BIl$$

and

torque per coil side $= TBIl \times d/2$

then

total torque $= 2TBIl\cdot d/2 = TBIld$

But T, the number of turns on the coil, B, the flux density in teslas and the dimensions of the coil, l and d, are all constant, and so

$$\text{torque} \propto I$$

Hence the p.m.m.c. instrument has a linear scale.

Example 11.2

A p.m.m.c. instrument is calibrated for a f.s.d. of 5 A and the control springs exert a full scale torque of 110 µNm. If the arc of the scale subtends an angle of 120°, determine the torque/rad exerted by the springs. What will be the torque when the scale reads 3·6 A?

$$\text{Angle subtended by scale} = 120°$$

$$= \frac{\pi}{180} \times 120 \text{ rad}$$

$$= 2\pi/3 \text{ rad}$$

But

total torque $= 110\,\mu\text{Nm}$

and so

torque/rad $= 110 \times \dfrac{3}{2\pi}\,\mu\text{Nm/rad}$

$$= 52\cdot5\,\mu\text{Nm/rad}$$

Since the instrument has a linear scale,

angle turned through for current of 3·6 A $= \dfrac{2\pi}{3} \times \dfrac{3\cdot6}{5} \text{ rad} = 0\cdot48\pi \text{ rad}$

torque produced $= 52\cdot5 \times 0\cdot48\pi\,\mu\text{Nm} = 79\,\mu\text{Nm}$

The torque exerted by the springs per radian deflection is 52·5 µNm, and for a current of 3·6 A the torque is 79 µNm.

Example 11.3

A p.m.m.c. voltmeter has a coil of 50 turns, having a resistance of 50 Ω. If the instrument gives full scale deflection for a current of 20 mA, how may it be adapted to read up to 300 V?

$$\text{Volt drop across coil} = I \times R_c$$

$$= 20 \times 10^{-3} \times 50\,\text{V} = 1\,\text{V}$$

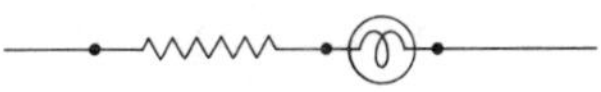

Fig. 11.1 Circuit for extension of voltmeter scale—Example 11.3.

For the instrument to be used for voltages up to 300 V, the voltage in excess of that required to produce f.s.d. must be dropped across a resistor connected in series with the coil. Fig. 11.1 shows the circuit of the resistor in series with the coil.

Voltage to be dropped across resistor = 300 − 1 V = 299 V

But   current in resistor = 20 mA   (series circuit)

Hence   resistance required $= \dfrac{299}{20 \times 10^{-3}} \, \Omega = 14{\cdot}95 \text{ k}\Omega$

Alternative solution. The current through the coil and the resistor is 20 mA, and so to drop 300 V,

$$\text{total resistance} = \dfrac{300}{20 \times 10^{-3}} \, \Omega = 15 \text{ k}\Omega$$

But the resistance of the coil is 50 Ω; therefore

$$\text{resistance to be added} = 15000 - 50 \, \Omega = 14{\cdot}95 \text{ k}\Omega$$

To adapt the instrument to read up to 300 V a 14·95 kΩ resistance must be connected in series.

Example 11.4
A p.m.m.c. galvanometer, having a coil of resistance 5 Ω, has to be adapted for use as an ammeter reading up to 15 A. How may this be done? If the galvanometer gives f.s.d. for a current of 15 mA, determine the value of the extra resistance required.

Since only 15 mA of the total 15 A has to pass through the coil, the re-

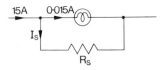

Fig. 11.2 Circuit and current distribution for extension of ammeter scale—Example 11.4.

mainder must be by-passed through a low resistance—a shunt—connected across the coil. Fig. 11.2 shows the circuit and the current distribution.

$$\text{Shunt current } I_S = 15 - 0{\cdot}015 \text{ A} = 14{\cdot}985 \text{ A}$$

Y

$$\text{Volt drop across the coil} = 15 \times 10^{-3} \times 5\,\text{V} = 75\,\text{mV}$$

Hence $\quad$ volt drop across shunt $= 75\,\text{mV} \quad$ (parallel circuit)

and $\quad$ resistance of the shunt $R_s = \dfrac{75 \times 10^{-3}}{14\cdot985}\,\Omega$

$$= 5\cdot005 \times 10^{-3}\,\Omega$$

The value of the shunt resistance required is $5\cdot005$ mΩ.

Example 11.5
The coil of a p.m.m.c. instrument is wound with 25 turns and f.s.d. is produced by a current of 40 mA. If the flux density produced by the permanent magnet is 0·1 T and if the control springs exert a full scale torque of 60 µNm, determine the area of the coil.

Example 11.6
The coil of a p.m.m.c. voltmeter is wound with 50 turns having a resistance of 50 Ω. If the radial flux density is 0·15 T and the coil measures 35 mm × 25 mm, determine the f.s.d. torque produced by the springs. For f.s.d. the instrument requires a magnetomotive force of 1 A.

Example 11.7
The scale of a cirscale voltmeter extends over an angle of $4\pi/3$ rad and the control springs exert a f.s.d. torque of 40 µNm for a reading of 300 V. What will be the reading of the voltmeter, and what will be the torque exerted by the springs when the pointer has moved through an angle of 80°?

Example 11.8
The coil of a p.m.m.c. ammeter has an area of 1200 mm^2 and consists of 15 turns. The radial density is 0·12 T and f.s.d. is produced by a coil current of 75 mA. Calculate the torque produced by the springs for f.s.d. If the range of the instrument extends over 120°, determine the work done against the springs for f.s.d.

Example 11.9
The coil of a p.m.m.c. ammeter has a resistance of 1 Ω and gives a f.s.d. for a current of 0·05 A. To what value would the range of the instrument be extended if a shunt resistance of 0·004 Ω were used in conjunction with the instrument?

Example 11.10
A p.m.m.c. voltmeter has a 12·46 kΩ resistor in series with the coil of 40 Ω resistance. If the full scale reading is 250 V, calculate the current required for f.s.d. If a section of the series resistor, equivalent to a resistance of 500 Ω is shorted out, will the instrument read high or low and by what percentage?

11.2 Moving Iron Instruments

There are two types of moving iron instrument (a) attraction and (b) repulsion. The control is either by springs or by gravity.

As movement of the iron in the instrument takes place, the inductance of the system varies. If the ratio between the change in inductance and the change in angular deflection is constant, then

$$\text{torque, } T \propto I^2$$

But if this ratio is not constant then

$$T = \tfrac{1}{2}I^2 \times \frac{L_2 - L_1}{\theta_2 - \theta_1}$$

Example 11.11
The f.s.d. torque in a moving-iron ammeter of the attraction type is 20 μNm for a maximum current of 5 A. What torque will be produced by a current of 3 A if the change in the inductance of the system is proportional to the change in the deflection?
For f.s.d.,

$$T = 20 \times 10^{-6} \, \text{Nm}$$
$$= k \times \tfrac{1}{2}I^2 \, \text{Nm}$$
$$= k \times 0\cdot5 \times 5^2 \, \text{Nm}$$

and so
$$k = \frac{20 \times 10^{-6}}{12\cdot5}$$

Hence for a current of 3 A,

$$T = \frac{20 \times 10^{-6}}{12\cdot5} \times 0\cdot5 \times 3^2 \, \text{Nm} = 7\cdot2 \, \mu\text{Nm}$$

A current of 3 A will produce a torque of 7·2 μNm.

Example 11.12
A moving-iron instrument requires, for f.s.d., a current of 0·02 A. If the average change of inductance with deflection is 2·5 mH/rad, estimate the torque for a f.s.d. of 120°.

$$T = \tfrac{1}{2}I^2 \times \text{change of inductance/rad}$$

$$= 0\cdot5 \times 0\cdot02^2 \times 2\cdot5 \times 10^{-3} \times \frac{2\pi}{3} \, \text{Nm}$$

$$= 19\cdot1 \, \mu\text{Nm}$$

The torque for f.s.d. is 19·1 μNm.

Example 11.13

A moving-iron voltmeter has a resistance of 20 Ω/V. If the instrument requires a magnetomotive force of 300 A for f.s.d., estimate the number of turns on the magnetizing coil.

Since the resistance of the voltmeter is 20 Ω/V, then

$$\text{current taken by meter} = \frac{1}{20}A = 0 \cdot 05 \text{ A}$$

But   magnetomotive force = current $\times$ number of turns

and so $$\text{number of turns} = \frac{300}{0 \cdot 05} = 6000$$

The magnetizing coil must be wound with 6000 turns.

Example 11.14

In a moving-iron instrument, a restoring torque of 18 μNm is balanced by that produced by a current of 2·5 A. What current is required to produce a torque of 15 μNm? The change in inductance of the system is proportional to the change in deflection.

Example 11.15

The variation of the inductance of a moving iron voltmeter, with deflection is given below:

Inductance in mH	545	555	560	575
Deflection in rad	0·4	0·6	0·72	1·06

Estimate the torque produced by a current of 10 mA if such a current causes a deflection of 0·5 rad. (Produce the curve backwards to obtain the inductance for zero deflection.)

Example 11.16

The coil of a moving-iron instrument is wound with 12 turns and has a resistance of 0·01 Ω. If the instrument requires a magnetomotive force of 300 A for f.s.d., determine the volt drop across the coil when it carries the current for f.s.d.

11.3 Rectifier Instruments

These instruments measure the average d.c. output of the rectifier system but the scale is graduated in r.m.s. values for wave forms having a form factor of 1·11.

Example 11.17

An alternating current, having an average value of $(2/\pi) \times$ the maximum value and a form factor of 1·08, is measured by a rectifier instrument. If the instrument reading is 3·5 A, estimate the true r.m.s. value of the current.

The rectifier instrument reads 1·11 × average value of sine wave. But the average value of the wave form is the same as that of a sine wave. Therefore

$$\text{average value of current} = \frac{3\cdot5}{1\cdot11}\,A = 3\cdot16\,A$$

and so $\qquad$ r.m.s. value of current $= 1\cdot08 \times 3\cdot16\,A = 3\cdot41\,A$

The r.m.s. value of the current is 3·41 A.

Example 11.18
A non-sinusoidal alternating current has an average value of 3·58 A and a form factor of 1·14. By what amount would a rectifier instrument in the circuit read high or low?

Example 11.19
A rectifier instrument measures an alternating voltage as 225 V. If this reading is known to be 5 V low, determine the form factor of the alternating voltage, assuming that it has the same average value as that of a sinusoidal voltage of the same frequency.

11.4 Thermal Instruments

In a closed circuit consisting of two dissimilar metals, current will flow if the junctions of the two metals are at different temperatures.

Fig. 11.3 shows a diagram of the thermal instrument with A and B being conductors of different metals. Since the temperature of the hot junction depends on the heat produced by the current, the micro-ammeter will read the r.m.s. values of the alternating currents. As the heating effect of a current is independent of frequency then thermal instruments may be used for measuring the values of high frequency currents.

Example 11.20
A moving-coil instrument and a thermal instrument are connected in series in a circuit in which a current in the form of $i = 5 \sin \omega t$ A is superimposed on a d.c. current of 4 A. Determine the readings on each of the two instruments.

The moving-coil instrument reads the average value. Since the average value of the alternating current is zero then the reading on the moving coil instrument will be equal to the direct current only, i.e. 4 A.

The thermal instrument will read the r.m.s. value and will depend on the heating effect of both the d.c. and the a.c.

Reading on thermal instrument $= \sqrt{(I_{d.c.}^2 + I_{a.c.}^2)}$
But $\qquad\qquad\qquad I_{a.c.} = I_m/\sqrt{2}$

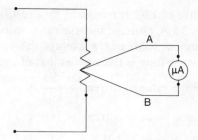

Fig. 11.3 Diagrammatic representation of a thermocouple instrument.

and so   instrument reading $= \sqrt{(4^2 + 5^2/2)}$ A

$= \sqrt{(16 + 12\cdot5)}$ A

$= \sqrt{28\cdot5}$ A $= 5\cdot34$ A

The moving-coil instrument will read 4 A and the thermal instrument will read 5·34 A.

Note: The r.m.s. value of a sum of alternating currents of sinusoidal form and of different frequencies is equal to the square root of the sum of the squares of the separate r.m.s. values.

$$I_t = \sqrt{(I_1^2 + I_2^2 + I_3^2 + \ldots)}$$

Example 11.21
A current of the form $i = 10 \sin 314t$ is maintained in a half-wave rectifier circuit, and the current from the rectifier is passed through a p.m.m.c. ammeter and a thermal ammeter. What will be the readings on the two ammeters?
If the rectifier circuit was for full wave rectification, what then would be the ammeter readings?

Example 11.22
What is the reading on a thermal instrument connected to measure the resultant current of $i_1 = 5 \sin 314t$ A and $i_2 = 3 \sin 942t$ A?

11.5 Instrument Errors

The error of an instrument reading is the difference between the indicated value and the true value. If the indicated value is greater than the true value the instrument is reading high and the error is positive, i.e. it requires a negative correction to give the true value.

The correction factor may be given as a percentage, i.e. ±2 per cent, or as a per unit (p.u.) value, i.e. $\pm0\cdot02$. This factor is usually stated as the accuracy of the instrument and is usually based on f.s.d. or over a specified section of the scale.

Example 11.23
A known a.c. voltage of 250 V is measured on a moving-iron instrument as 254 V. Determine the error and the p.u. correction factor to be applied.

$$\text{Error} = \text{indicated value} - \text{true value}$$
$$= 254-250 \text{ V} = 4 \text{ V}$$
$$\text{Correction factor} = \frac{-\text{error}}{\text{true value}} \text{ p.u.}$$
$$= \frac{-4}{250}\text{p.u.} = -0\cdot016 \text{ p.u.}$$

At 250 V the instrument is reading 4 V high and the required correction factor is $-0\cdot016$ p.u.

Example 11.24
An ammeter in a d.c. circuit reads 2·475 A. When the same current was passed through two parallel conductors placed 100 mm apart the force between the conductors was determined as 12·5 µN/m. Calculate the true value of the current, the error and the p.u. correction factor for the ammeter.
    The force between parallel conductors carrying the same current is given by

$$F = \frac{0\cdot2 \times 10^{-6} \times I^2 \times l}{d}$$

where l is the length of the conductors in metres and d is the distance, in metres, between the centres of the conductors. Thus

$$12\cdot5 \times 10^{-6} = \frac{0\cdot2 \times 10^{-6} \times I^2 \times 1}{0\cdot1}$$
$$12\cdot5 = 2I^2$$
$$I^2 = 6\cdot25$$
$$I = 2\cdot5 \text{ A}$$
$$\text{Error} = \text{indicated value} - \text{true value}$$
$$= 2\cdot475 - 2\cdot5 \text{ A} = -0\cdot025 \text{ A}$$
$$\text{p.u. correction factor} = \frac{+0\cdot025}{2\cdot5} = +0\cdot01$$

The true value of the current is 2·5 A; hence the ammeter is reading low with an error of 0·025 A and the p.u. correction factor is $+0\cdot01$.

Example 11.25

A cathode ray oscilloscope, with a calibration constant of 10 V/mm, was used to check the accuracy of a moving-iron voltmeter. When the voltmeter gave a reading of 225 V a trace on the oscilloscope showed a distance of 65·0 mm between peaks. Determine the error in the instrument reading and the correction factor which should be applied.

Example 11.26

An ammeter is known to require a p.u. correction factor of ± 0.018. If the ammeter measures a current as 4·91 A, calculate the true value of the current.

Answers

Chapter 1

1.3	-0.255 A; 0.292 A; -0.036 A
1.4	$I_A = 7.15$ A; $I_B = 11.42$ A
1.5	4 A; 8
1.6	$_cI_B = 20$ A; $_AI_C = 0$; $_DI_C = 20$ A; $_cI_A = 1.74$ A
1.7	$I_{30} = 1.3$ A; $I_{20} = 0.66$ A; $I = 1.96$ A
1.8	$I_A = 8.57$A; $I_B = -9.62$ A; $I = 1.06$ A
1.9	1.9 A; 2.25 A
1.14	200 A; 150 A; 80 A; 0 A. 235 V; 230 V; 229 V
1.15	1367 m or 1.367 km
1.16	237 V; 235 V; 237 V
1.17	71.5 A; 3.5 A; 103.5 A. 236 V; 236 V
1.18	78.3 A
1.19	233 V; 231 V
1.20	234 V; 232 V; -239 V
1.21	16 A; -34 A. 6 A; -24 A; 26 A; -14 A. -22 A; 8 A; 32 A. 233 V; 233 V; 234 V
1.24	(a) 0.555 A; 0.075 A; 0.63 A. (b) 0.688 A; -0.125 A; -1.0 A. 0.813 A; 0.875 A
1.25	0.78 A; 2.3 A; 3.08 A

Chapter 2

2.4	111 kΩ; 0.01 Ω
2.5	6 Ω; 0.16 A
2.6	9.23 m
2.7	6 mA
2.8	32.3°C
2.12	8 V; 0.8 A
2.13	0.33 A
2.14	1.52%; 761.4 mm
2.15	11.9 Ω
2.16	1.245 mA
2.17	12.66 Ω
2.19	7.69 Ω
2.20	80 Ω
2.21	4.732 kΩ

Chapter 3

3.3	40 μT
3.5	40 mm
3.6	30 A
3.7	$\pm 0.167\%$
3.12	220 mm
3.13	0.833 T
3.14	0.08 N/mm
3.15	9.95×10^{15} electrons/s
3.16	4.78 kN/m^2
3.17	45 mm; 0 N
3.18	24.5 A; 0.8 mN
3.22	27 μNm
3.23	5.33 mT

3.24	1·04 A
3.25	0·925 T
3·26	16 A; 250 V
3.27	85·7 rad/s
3.32	2·4 A
3.33	5 kA/m; 6·3 mT
3.34	$0·6 \times 10^9$ units: 144 kA
3.35	885 A
3.36	480 μWb

Chapter 4

4.4	200×10^3 SI units
4.5	1·06 kA
4.6	$1·2 \times 10^6$ SI units
4.7	$1·5 \times 10^6$ SI units; 0·6 mWb; 0·4 mWb
4.8	608 A
4.9	2·63 A
4.10	1·28 kA
4.11	0·525 mWb
4.12	1800
4.13	16 turns
4.16	1·15 T
4.17	78 A
4.18	86 A/m
4.19	2·2 kA
4.23	3·3 A
4.24	3·84 A
4.25	955; 0·88 T
4.26	390 μWb
4.27	770
4.28	1·12
4.29	0·79 m

Chapter 5

5.6	1·2 V
5.7	0·59 T
5.8	874 mm/s
5.9	1·3 mV
5.10	1·8 N; 2·7 W; 0·54 V
5.11	240 V
5.12	150 rad/s
5.17	1·25 H
5.18	0·2 s

5.19	6 A
5.20	0·2 H
5.21	1·02 mH
5.22	0·1 s
5.23	3·14 mH; 15·7 mV
5.24	H^{-1}
5.27	0·01 s
5.28	40 Ω
5.29	3·7 A; 6·2 A
5.30	88 V; 2·2 A
5.31	20 A; 15·5 A
5.32	53·3 V; $-13·3$ V; 43·3 V
5.38	0·4 H
5.39	25 A/s
5.40	1 H
5.41	30×10^3 H^{-1}
5.42	0·5 H
5.43	0·72 H; 0·6 H; 0·83
5.44	8 V
5.45	413 turns

Chapter 6

6.4	2·5 C
6.5	$+30$ V; -10 V
6.6	40 kV/m; 100 V
6.8	2 mm
6.11	1·59 mC/m^2; 398 μC/m^2
6.12	500 mm^2
6.13	160 μC
6.17	1·42 μC/m^2; 2·84 μC/m^2
6.18	$11·3 \times 10^{-3}$ m^2
6.19	236 V
6.20	3·52
6.25	2·5 V
6.26	133 pF
6.27	2·36 mm
6.28	222 μC/m^2
6.29	4·95
6.30	200 V; 700 V
6.31	2 nF; 1·67 nF
6.35	185 V; 200 μC
6.36	1·84 mC
6.37	6 μF
6.38	10 s

6.39   104 V

6.40   3·75 MΩ in parallel

6.41   758 μC

6.42   18·3 s

6.43   30·5 s; 33·5 s

Chapter 7

7.3   2 μF

7.4   375 μC; 6 μF

7.5   10 s

7.6   320 μC; 107 V; 80 V; 53 V

7.7   192 V; 38 V

7.10   1·2 mC; 2·4 mC; 3·6 mC; 36 μF

7.11   9 μF

7.12   137 V

7.13   180 V

7.14   12 μF; 100 V

7.16   3 μF; 1·5 μF; 20 μF; 4 μF

7.17   1, 1·6, 2·4, 3, 4, 5·3, 6·7, 10, 16 μF

7.18   10 parallel rows of 4 in series

7.19   6 μF

7.20   77 V; 38 V; 77 V; 58 V

7.24   5

7.25   3·75 mm

7.26   6·75 mm

7.27   64·5 pF

7.28   870 pC; 102·5 V

7.29   124 pF

7.30   88·5 pF

Chapter 8

8.6   1000 rev/min

8.7   16·7 Hz

8.8   45 Hz; 250 Hz

8.9   104·7 rad/s; 6284 rad/s

8.10   0·06 s; 1 ms

8.11   166 V; 1·4 ms

8.12   1·89 ms; 8·11 ms; 1·6 A

8.15   5·88 A

8.16   157 V

8.17   70 V

8.18   1·6 Wb/s

8.19   60 V

8.20   40 V

8.21   25 V

8.25 $v = 60 \sin 300t$; 42·4 V

8.26   342 V; 242 V

8.27   8·6 A

8.28   4·6 A

8.29   Flat topped; peaky

8.30 $i = 8·6 \sin(314t - 16°)$ A

8.33   5·66 A; 3·6 A

8.34   444 V

8.35   1·145

8.36   1·09

Chapter 9

9.2 $v = 150 \sin 314t$ V

9.3   20 Ω

9.4 $v = 340 \sin 314t$ V
       $i = 8·5 \sin 314t$ A

9.5   20 A

9.8   95·5 mH

9.9 $v = 665 \sin(314t + \pi/2)$ V

9.10   6·7 A

9.11   45 Hz

9.12   780 turns

9.16   63·6 Hz

9.17 $i = 4·7 \sin(110\pi t + \pi/2)$ A

9.18   47·2 μF

9.19   76·5 μF; 7·2 A

9.20 $v = 338 \sin(314t - \pi/2)$ V; 159 V; 79 V

9.24 $v = 170 \sin(314t + 37°)$ V

9.25   10 A; 67·5° (lag)

9.26   8 A; 37° (lag); 161 V

9.27   3·33 A; 53° (lag); 122 V; 155 V

9.28   102 mH

9.29   19·2 Ω; 46 mH

9.30   25·8 Ω; 48·5 mH

9.31   0·21 H

9.33   5·93 A; 51·8° (lead)

9.34   73 μF

9.35   175 μF

9.36   6·13 Ω; 5·14 Ω; 43 A

9.37　35·4 V; 53° (lead)

9.42　3·84 kW

9.43　2·89 kW

9.44　9 Ω; 38·3 mH

9.45　2·5 kW; 3·125 kVA; 24 mH

9.46　8·8 kW; 51 A; 12·22 kVA;
　　　5·32 kVAr

9.47　52 Ω; 325 W; 152 W

Chapter 10

10.5　6·53 kΩ; 6·15 kΩ; 120 V

10.6　155 × 10^{15} electrons/s

10.7　13·9 mA; 0·335 A

10.8　4·25 kΩ; 3·21 kΩ

10.9　22 mA; 82 V

10.10　86·7 ms

10.12　10·35 mW; 364 kΩ

10.13　11·8 V; 12·2 V

10.14　4·46 V; 4·95 V

10.15　4·62 μA; 135 V

Chapter 11

11.5　600 mm^2

11.6　131 μNm

11.7　100 V; 13·3 μNm

11.8　162 μNm; 339 μJ

11.9　12·55 A

11.10　20 mA; 4·18% (high)

11.14　2·28 A

11.15　3 μNm

11.16　0·25 V

11.18　107 mA (low)

11.19　1·135

11.21　3·18 A; 5·0 A; 6·36 A; 7·1 A

11.22　4·12 A

11.25　+2·17% or +0·0217 p.u.

11.26　5 A